Two-Dimensional Semiconductors

Two-Dimensional Semiconductors

Synthesis, Physical Properties and Applications

Jingbo Li
Zhongming Wei
Jun Kang

Authors

Prof. Jingbo Li
South China Normal University
Institute of Semiconductors
No.55, West of Zhongshan Avenue
510631 Guangzhou
China

Prof. Zhongming Wei
Institute of Semiconductors, CAS
State Key Laboratory of Superlattices
and Microstructures
No.A35, QingHua East Road
Haidian District
100083 Beijing
China

Prof. Jun Kang
Beijing Computational Science Research
Center
Materials and Energy Division
Building 9, No.10 Xibeiwang East Road
Haidian District
100193 Beijing
China

Cover
Cover Image: © Rost9/Shutterstock

All books published by **Wiley-VCH** are carefully produced. Nevertheless, authors, editors, and publisher do not warrant the information contained in these books, including this book, to be free of errors. Readers are advised to keep in mind that statements, data, illustrations, procedural details or other items may inadvertently be inaccurate.

Library of Congress Card No.:
applied for

British Library Cataloguing-in-Publication Data
A catalogue record for this book is available from the British Library.

Bibliographic information published by the Deutsche Nationalbibliothek
The Deutsche Nationalbibliothek lists this publication in the Deutsche Nationalbibliografie; detailed bibliographic data are available on the Internet at <http://dnb.d-nb.de>.

© 2020 Wiley-VCH Verlag GmbH & Co. KGaA, Boschstr. 12, 69469 Weinheim, Germany

All rights reserved (including those of translation into other languages). No part of this book may be reproduced in any form – by photoprinting, microfilm, or any other means – nor transmitted or translated into a machine language without written permission from the publishers. Registered names, trademarks, etc. used in this book, even when not specifically marked as such, are not to be considered unprotected by law.

Print ISBN: 978-3-527-34496-3
ePDF ISBN: 978-3-527-81593-7
ePub ISBN: 978-3-527-81595-1
oBook ISBN: 978-3-527-81596-8

Typesetting   SPi Global, Chennai, India
Printing and Binding   CPI books GmbH, Leck

Printed on acid-free paper

10 9 8 7 6 5 4 3 2 1

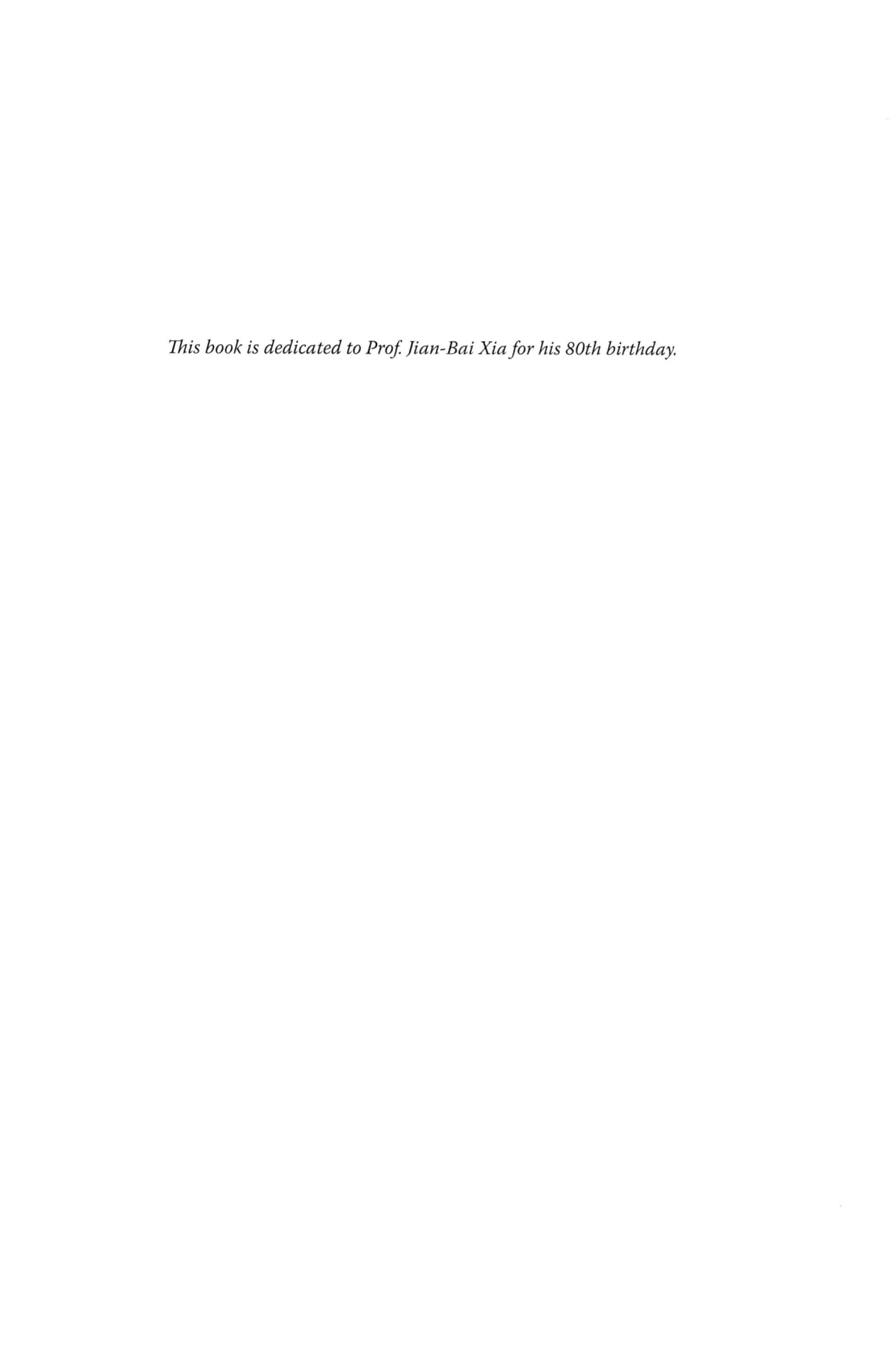

This book is dedicated to Prof. Jian-Bai Xia for his 80th birthday.

Preface

Ultrathin two-dimensional (2D) materials, such as graphene and MoS_2, have attracted broad interest because of their exotic condensed-matter phenomena that are absent in bulk counterparts. Graphene, which is composed of a single layer of carbon atoms arranged in honeycomb lattice, has a linear dispersion near the K point, and charge carriers can be described as massless Dirac fermions, providing abundant physical picture. In contrast, 2D transition metal dichalcogenides (TMDs), transition metal oxides, black phosphorus, and boron nitride (BN) exhibit versatile optical, electronic, catalytic, and mechanical properties. It was reported that the 2D materials, especially 2D semiconductors with the intrinsic nanometer-scale size, can help to extend Moore's law, which face the challenge of further scaling down the transistor channel.

In this book, we discuss the theoretical study, synthetic method, the unique properties, the potential application, the challenges, and opportunities of the 2D semiconductors. Firstly, a general introduction of 2D materials was given. Then, the theoretical study including electronic structures and predications, the reparation, properties and applications in (opto) electronics or other devices of 2D materials/semiconductors and their alloys, and heterostructures were discussed in detail. At last, a perspective and outlook of this fast developing field is summarized.

I became a PhD student under the supervision of Prof. Jian-Bai Xia from 1998 to 2001. He gave plenty of guidelines to my study, research, and my life. We acknowledge him by dedicating this book on his 80th birthday this year. In addition, we convey our best wishes to him and his family and also to his fruitful research and healthy and happy life.

We sincerely hope this book can help researchers to understand 2D materials.

Beijing *Jingbo Li*
10 December 2019

Contents

About the Authors *xiii*
Acknowledgments *xv*

1 Introduction *1*
1.1 Background *1*
1.2 Types of 2D Materials *4*
1.3 Perspective of 2D Materials *6*
References *7*

2 Electronic Structure of 2D Semiconducting Atomic Crystals *9*
2.1 Theoretical Methods for Study of 2D Semiconductors *9*
2.1.1 Density Functional Theory *9*
2.1.2 Linear Scaling Three-Dimensional Fragment (LS3DF) Method *10*
2.1.3 GW Approximation *10*
2.1.4 Semiempirical Tight-Binding Method *10*
2.1.5 Nonequilibrium Green's Function Method *11*
2.2 Electronic Structure of 2D Semiconductors *11*
2.2.1 Graphyne Family Members *11*
2.2.2 Nitrogenated Holey Graphene *14*
2.2.3 Transition Metal Dichalcogenides *15*
2.3 Prediction of Novel Properties in 2D Moiré Heterostructures *19*
2.3.1 $MoS_2/MoSe_2$ Moiré Structure *19*
2.3.2 Graphene/Nitrogenated Holey Graphene Moiré Structure *26*
2.3.2.1 Atomic Structure: Ordered Stacking Versus Moiré Pattern *26*
2.3.2.2 Renormalized Fermi Velocity *31*
References *33*

3 Tuning the Electronic Properties of 2D Materials by Size Control, Strain Engineering, and Electric Field Modulation *35*
3.1 Size Control *35*
3.2 Strain Engineering *40*
3.3 Electric Field Modulation *48*
References *52*

4	**Transport Properties of Two-Dimensional Materials: Theoretical Studies** *55*
4.1	Symmetry-Dependent Spin Transport Properties of Graphene-like Nanoribbons *55*
4.1.1	Graphene Nanoribbons *55*
4.1.2	Graphyne Nanoribbon *57*
4.1.3	Silicene Nanoribbons *59*
4.2	Charge Transport Properties of Two-Dimensional Materials *61*
4.2.1	Phonon Scattering Mechanism in Transport Properties of Graphene *61*
4.2.2	Phonon Scattering Mechanism in Transport Properties of Transition Metal Dichalcogenides *63*
4.2.3	Anisotropic Transport Properties of 2D Group-VA Semiconductors *67*
4.3	Contacts Between 2D Semiconductors and Metal Electrodes *69*
4.3.1	Carrier Schottky Barriers at the Interfaces Between 2D Semiconductors and Metal Electrodes *69*
4.3.2	Partial Fermi Level Pinning and Tunability of Schottky Barrier at 2D Semiconductor–Metal Interfaces *70*
4.3.3	Role of Defects in Enhanced Fermi Level Pinning in 2D Semiconductor/Metal Contacts *72*
	References *75*

5	**Preparation and Properties of 2D Semiconductors** *79*
5.1	Preparation Methods *79*
5.1.1	Mechanical Exfoliation *79*
5.1.2	Liquid-Phase Exfoliation *81*
5.1.3	Vapor-Phase Deposition Techniques *85*
5.2	Characterizations of 2D Semiconductors *90*
5.2.1	Surface Morphology (SEM, OM, and TEM) *90*
5.2.2	Thickness (Raman, AFM, and HRTEM) *92*
5.2.3	Phase Structure (HRTEM and STEM) *93*
5.2.4	Band Structure (Optical Absorption and Photoluminescence, ARPES) *94*
5.2.5	Chemical Composition and Chemical States (XPS and EDS) *94*
5.3	Electrochemical Properties of 2D Semiconductors *96*
	References *97*

6	**Properties of 2D Alloying and Doping** *99*
6.1	Introduction *99*
6.2	Advantages of 2D Alloys *99*
6.2.1	Adjustable Bandgap *100*
6.2.2	Carrier-Type Modulation *103*
6.2.3	Phase Change *104*
6.2.4	Application of 2D Semiconductor Alloys in the Field of Magnetism *107*
6.2.5	Improve Device Performance *108*

6.3	Preparation Methods for 2D Alloys	*110*
6.3.1	Chemical Vapor Transport (CVT)	*110*
6.3.2	Physical Vapor Deposition (PVD)	*111*
6.3.3	Chemical Vapor Deposition (CVD)	*113*
6.4	Characterizations of 2D Alloys	*114*
6.4.1	STEM	*115*
6.4.2	Raman Spectroscopy	*115*
6.4.3	Photoluminescence (PL) Spectrum	*119*
6.5	Doping of 2D Semiconductors	*119*
	References	*121*
7	**Properties of 2D Heterostructures**	*123*
7.1	Conception and Categories of 2D Heterostructures	*123*
7.2	Advantages and Application of 2D Heterostructures	*125*
7.3	Preparation Methods for 2D Heterostructures	*129*
7.3.1	Mechanical Transfer: Liquid Method and Dry Method	*130*
7.3.2	Chemical Methods	*131*
7.4	Characterizations of 2D Heterostructures	*137*
	References	*139*
8	**Application in (Opto) Electronics**	*143*
8.1	Field-Effect Transistors	*143*
8.2	Infrared Photodetectors	*145*
8.2.1	Figures of Merit	*146*
8.2.2	Photodetection Mechanism	*147*
8.2.2.1	Photothermoelectric Effect	*147*
8.2.2.2	Bolometric Effect	*147*
8.2.2.3	Photogating Effect	*148*
8.2.2.4	Photovoltaic Effect	*148*
8.2.2.5	Plasmonic Effect	*148*
8.2.3	Typical 2D-Based Infrared Photodetectors	*149*
8.2.3.1	Graphene Infrared Photodetectors	*149*
8.3	2D Photodetectors with Sensitizers	*151*
8.3.1	Graphene-based Hybrids Detectors	*151*
8.3.2	TMD-Based Hybrid Detectors	*152*
8.3.3	Plasmonic Sensitized Detectors	*153*
8.4	New Infrared Photodetectors with Narrow Bandgap 2D Semiconductors	*155*
8.5	Future Outlook	*156*
8.5.1	Optoelectronic Memory of 2D Semiconductors	*156*
8.5.2	Solar Cells	*161*
	References	*162*
9	**Perspective and Outlook**	*165*
	Index	*167*

About the Authors

Jingbo Li received his PhD degree from the Institute of Semiconductors, Chinese Academy of Sciences, in 2001 under the supervision of Prof. Jian-Bai Xia. Then, he spent six years at the Lawrence Berkeley National Laboratory and National Renewable Energy Laboratory in USA. From 2007 to 2019, he worked as a professor at the Institute of Semiconductors, Chinese Academy of Sciences. Since 2019, he became a full-time professor and the dean of Institute of Semiconductors, South China Normal University. His research interests include the design, fabrication, and application of novel nanostructured semiconductors. He has published more than 290 scientific publications with more than 15000 citations.

Zhongming Wei received his BS degree from Wuhan University (China) in 2005 and PhD degree from the Institute of Chemistry, Chinese Academy of Sciences, in 2010 under the supervision of Prof. Daoben Zhu and Prof. Wei Xu. From August 2010 to January 2015, he worked as a postdoctoral fellow and then as an Assistant Professor in Prof. Thomas Bjørnholm's group at the University of Copenhagen, Denmark. Currently, he is working as a professor at the Institute of Semiconductors, Chinese Academy of Sciences. His research interests include low-dimensional semiconductors and their (opto)electronic devices.

Jun Kang received his PhD degree from the Institute of Semiconductors, Chinese Academy of Sciences, in 2014. After that, he performed his postdoctoral research at the University of Antwerp in Belgium and Lawrence Berkeley National Laboratory in USA. In 2019, he joined Beijing Computational Science Research Center as an assistant professor. His research field is first-principles calculations on novel electronic properties of low-dimensional semiconductors. He has published over 60 peer-reviewed articles with more than 4000 citations.

Acknowledgments

Finally, I would like to thank all my group members who spent a lot time for the writing and revising of this book: Dr. Bo Li, Dr. Mianzeng Zhong, Dr. Yan Li, Dr. Le Huang, Dr. Nengjie Huo, Dr. Xiaoting Wang, Ziqi Zhou, Jingzhi Fang, Kai Zhao, Yu Cui, and Longfei Pan. Without their hard work and contribution, we would not be able to finish this book on time. Thanks to Project Editor Ms. Shirly Samuel at Wiley-VCH for all her help in the publication of this book.

1

Introduction

1.1 Background

In 2004, Ander Geim and Konstantin Novoselov from the University of Manchester, UK, first obtained graphene sheets by mechanical exfoliation method, successfully fabricated the first graphene field effect transistor (FET), and investigated its unique physical properties [1]. Before the discovery of graphene, according to the thermodynamic fluctuation law, the two-dimensional (2D) atomic thick layer under nonabsolute zero degrees is unlikely to exist stably [2]. Why is graphene stable at temperatures above absolute zero? Further theoretical studies have shown that this is because large-scale graphene is not distributed in a perfect 2D plane but in a wave-like shape. The experimental results support this view [3, 4]. Therefore, the discovery of graphene shocked the condensed matter physics community and also quickly ignited the enthusiasm of scientists to study 2D materials (a crystalline material composed of a single atomic layer or few atomic layers), indicating the arrival of the "two-dimensional material era."

In 2010, Ander Geim and Konstantin Novoselov were awarded the Nobel Prize in Physics for their outstanding contribution to graphene (Figure 1.1) [1]. Graphene is a 2D material composed of carbon atoms and having a hexagonal lattice structure. Graphene has good toughness and its Young's modulus can theoretically reach as 1 TPa [5]. Therefore, graphene can form different structures through different curved stacks, such as zero-dimensional fullerenes, one-dimensional carbon nanotubes, and three-dimensional stacked graphite [6].

Graphene has shown many excellent physical properties resulting from the unique structure, and the disappearance of interlayer coupling makes the two carbon atoms in the cell completely equivalent, thus making the effective mass of electrons on the Fermi surface zero [7–11]. Because graphene has a unique Dirac band structure, carriers can completely tunnel in graphene, and electrons and holes in graphene have a very long free path. Therefore, the electronic transport of graphene is hardly affected by phonon collisions and temperature [8].

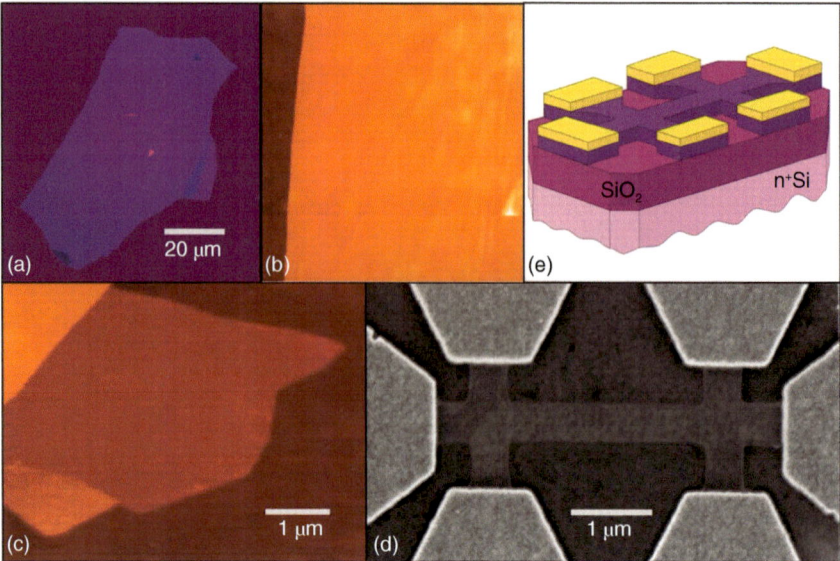

Figure 1.1 Graphene films. (a) Photograph (in normal white light) of a relatively large multilayer graphene flake with thickness ∼3 nm on top of an oxidized Si wafer. (b) Atomic force microscope (AFM) image of 2 μm by 2 μm area of this flake near its edge. Colors: dark brown, SiO_2 surface; orange, 3 nm height above the SiO_2 surface. (c) AFM image of single-layer graphene. Colors: dark brown, SiO_2 surface; brown-red (central area), 0.8 nm height; yellow-brown (bottom left), 1.2 nm; orange (top left), 2.5 nm. Notice the folded part of the film near the bottom, which exhibits a differential height of ∼0.4 nm. (d) SEM micrograph of an experimental device prepared from few-layer graphene, and (e) its schematic view. Source: Reproduced with permission from Novoselov et al. [1]. Copyright 2004, The American Association for the Advancement of Science.

The mobility of electrons in monolayer graphene is much larger than that in its parent graphite (Figure 1.2c) [16]. In addition, graphene has shown good thermal conductivity (Figure 1.2d) [17], room temperature quantum Hall effect (Figure 1.2a) [12, 14], single-molecule detection (Figure 1.2b), and high light transmission [18]. Graphene is a semimetal material without band gap, it cannot form a good switching ratio in terms of regulation, thus greatly limiting the application of graphene in electronic devices. Although on the bilayer and multilayer graphene, the graphene can obtain a certain band gap by applying an electric field and stress [19]. However, this band gap is not only small but also has a low electrical on/off ratio and is difficult to apply to a controllable device. With the extensive research on two-dimensional materials, it is found that the disadvantages of graphene are compensated for in other families of 2D materials [20–26].

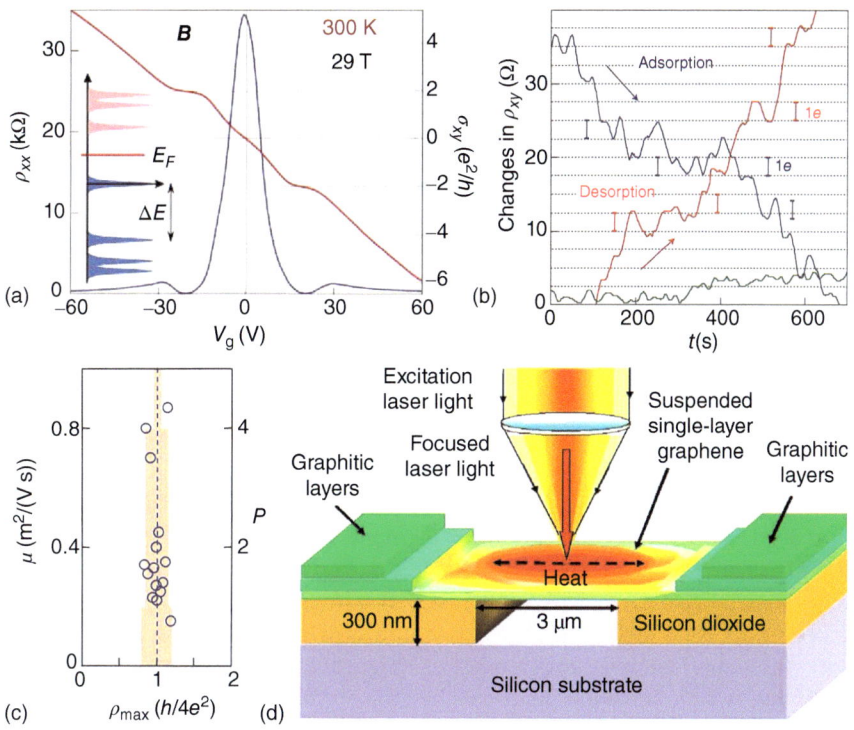

Figure 1.2 (a) Room temperature quantum Hall effect in graphene. σ_{xy} (red) and ρ_{xx} (blue) as a function of gate voltages (V_g) in a magnetic field of 29 T. The need for high B is attributed to broadened Landau levels caused by disorder, which reduces the activation energy. Source: Reproduced with permission from Novoselov et al. [12]. Copyright 2007, The American Association for the Advancement of Science. (b) Single-molecule detection in graphene. Examples of changes in Hall resistivity observed near the neutrality point ($|n| < 10^{11}$ cm^{-2}) during adsorption of strongly diluted NO_2 (blue curve) and its desorption in vacuum at 50 °C (red curve). The green curve is a reference – the same device thoroughly annealed and then exposed to pure He. The curves are for a three-layer device in $B = 10$ T. The adsorbed molecules change the local carrier concentration in graphene one by one electron, which leads to step-like changes in resistance. The achieved sensitivity is due to the fact that graphene is an exceptionally low-noise material electronically. Source: Reproduced with permission from Schedin et al. [13]. Copyright 2007, Nature Publishing Group. (c) Mobility of graphene. Maximum values of resistivity $\rho = 1/\sigma$ (circles) exhibited by devices with different mobilities μ (left y axis). The histogram (orange background) shows the number P of devices exhibiting ρ_{max} within 10% intervals around the average value of $\sim h/4e^2$. Several of the devices shown were made from two or three layers of graphene, indicating that the quantized minimum conductivity is a robust effect and does not require "ideal" graphene. Source: Reproduced with permission from Novoselov et al. [14]. Copyright 2005, Nature Publishing Group. (d) Schematic of the experiment showing the excitation laser light focused on a graphene layer suspended across a trench. The focused laser light creates a local hot spot and generates a heat wave inside single-layer graphene propagating toward heat sinks. Source: Reproduced with permission from Balandin et al. [15]. Copyright 2008, American Chemical Society.

1.2 Types of 2D Materials

The rapid pace of progress in graphene and the methodology developed in synthesizing ultrathin layers have led to exploration of other 2D materials, such as monolayer of group IVA elements (silicon, germanium, and tin) [27, 28] and their adjacent group elements (such as boron and phosphorus) monolayers; 2D layered metal oxides or metal hydroxides (octahedral or orthogonal tetrahedral structure in the layer) [29]; transition metal dichalcogenides (TMDCs) [21]; and graphene analogs such as boron nitride (BN) [30]. These 2D materials ranging from insulators (e.g. BN), semiconductors (e.g. TMDCs, tellurene, $PtSe_2$, and BP) to semimetals (e.g. $MoTe_2$), topological insulators (e.g. Bi_2Se_3), superconductors ($NbSe_2$), and metals (1T-VS_2) exhibit diverse property.

There are many 2D materials, and some literature studies have classified the 2D materials based on their positions in periodic table of elements (Figure 1.3a) [22], stoichiometric ratios [31], space groups, and structural similarities [32]. The advantage of classifying 2D materials by periodic table of elements is that 2D materials with the same group of elements often have similar properties, which has a good guiding significance for finding novel 2D materials. In 2017, Michael Ashton et al. have found that 826 2D materials can be grouped according to their stoichiometric ratios and 50% of the layered materials are represented by just five stoichiometries (Figure 1.3b) [31]. The advantage of classifying two-dimensional materials by stoichiometric ratios is to distinguish 2D materials with different stoichiometric ratios but the same elements. At the same time, when synthesizing 2D materials, the vapor pressure of growth can be adjusted according to the stoichiometric ratios, which is conductive to synthetic materials. Recently, Nicolas Mounet et al. developed a system based on high-throughput computational exfoliation of 2D materials (Figure 1.3c) [32]. They searched for materials with layered structure from more than 100 000 kinds of three-dimensional compounds in the existing database, and the 1036 kinds of easily exfoliable cases provide novel structural prototypes and simple ternary compounds by high-throughput calculations. They classify the 2D materials of the easily exfoliated group into different prototypes, according to their space groups and their structural similarities. The structure of 2D materials can be useful to search for more suitable substrates. 2D materials with a similar structure can often form stable 2D alloys. In this book, we will focus on the electronic structure, synthesis, and applications of 2D materials. We will classify 2D materials into three types based on the synthesis, structure, and application: 2D single, doped components, and van der Waals heterostructures.

2D single materials, such as graphene and MoS_2, generally refer to materials that can be exfoliated from corresponding van der Waal layered three-dimensional materials. 2D doped materials include adsorption, intercalation, substitution doping, and so on. In this book, we focus on the substitution doping: transition metal element or chalcogen element is substituted by other element, such as $MoS_{2(1-x)}Se_{2x}$ [33] and Fe-doped SnS_2 [34]. 2D heterostructures contain vertical and lateral types (Figure 1.4).

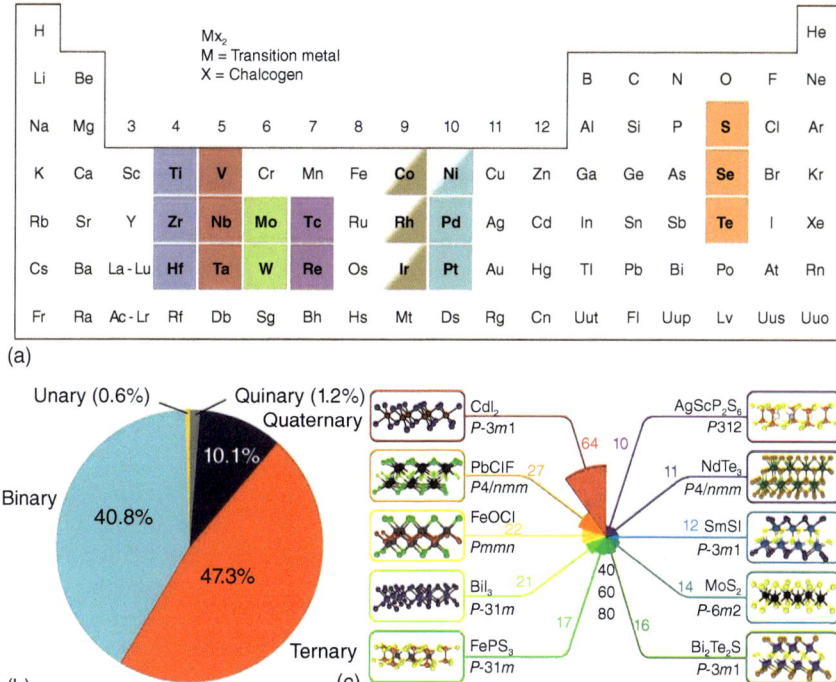

Figure 1.3 Classification of 2D materials. (a) About 40 different layered TMDCs compounds exist. The transition metals and the three chalcogen elements that predominantly crystallize in those layered structure are highlighted in the periodic table. Source: Reproduced with permission from Chhowalla et al. [22]. Copyright 2013, Nature Publishing Group. (b) Distribution of stoichiometries of the 826 layered compounds. Source: Reproduced with permission from Ashton et al. [31]. Copyright 2017, American Physical Society. (c) Polar histogram showing the number of structures belonging to the 10 most common 2D structural prototypes in the set of 1036 easily exfoliable 2D materials. A graphical representation of each prototype is shown, together with the structure-type formula and the space group of the 2D systems. The room temperature values of the thermal conductivity in the range $\sim(4.84 \pm 0.44) \times 10^3$ to $(5.30 \pm 0.48) \times 10^3$ W/mK were extracted for a single-layer graphene from the dependence of the Raman G peak frequency on the excitation laser power and independently measured G peak temperature coefficient. Source: Reproduced with permission from Mounet et al. [32]. Copyright 2018, Springer Nature.

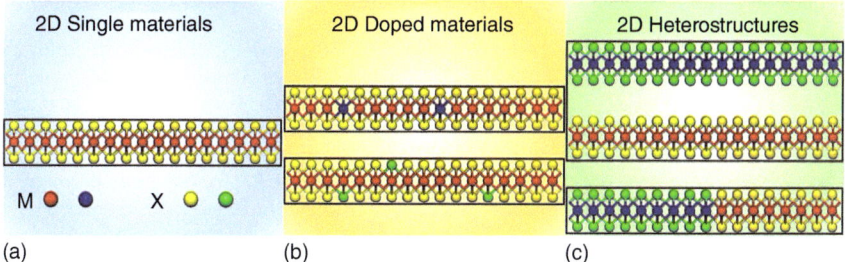

Figure 1.4 Atomic structure of (a) 2D single materials, (b) 2D doped materials, and (c) 2D heterostructures. Red and blue balls stand for transition metal element (M), yellow and green balls represent chalcogen element (X).

1.3 Perspective of 2D Materials

2D materials have been attracting wide interest because of their peculiar structural properties and fascinating applications in the areas of electronics, optics, magnetism, biology, and catalysis. Overall, the current research on 2D materials is mainly in two aspects: (i) Wafer-scale growth of 2D materials and their industrial applications. (ii) Synthesis of novel 2D materials and study their physicochemical properties.

The ability to grow large, high-quality single crystals for 2D components is essential for the industrial application of 2D devices. Until now, some 2D materials, such as MoS_2 (Figure 1.5a), WS_2 (Figure 1.5b), InSe (Figure 1.5c), and BN (Figure 1.5d), have been synthesized as wafer scale by vapor-phase deposition or pulsed laser deposition method. Thus, developing a simple and low-cost method to synthesize wafer-scale 2D materials is a current research focus. On the other hand, taking advantage of unique characteristics of 2D materials, direct integration based on 2D heterostructures is an ingenious method (Figure 1.5e) [39].

Although some 2D materials have been synthesized and investigated now, there are more than 1000 2D materials in theory and many of them still have a lot to discover, which are suggested to have peculiar property and need further study. The efforts on exploiting the application of 2D materials in optoelectronic and

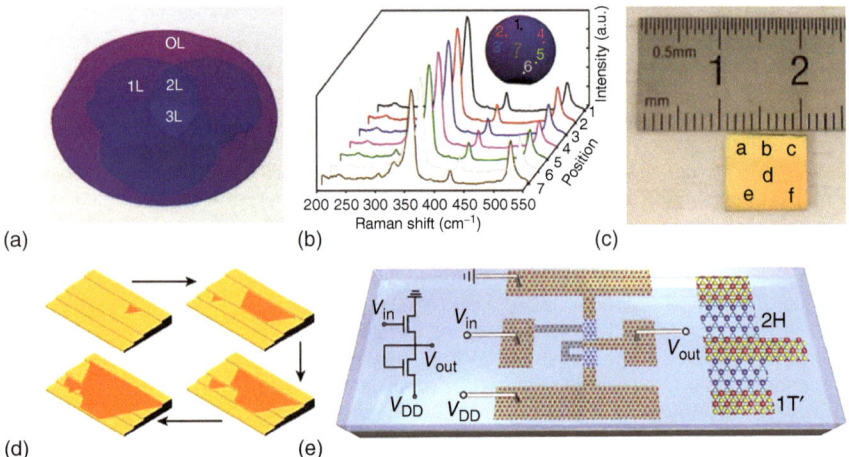

Figure 1.5 Wafer-scale growth and integrated circuit of 2D materials. (a) Three wafer-scale MoS_2 films transferred and stacked on a 4 in. SiO_2/Si wafer. Source: Reproduced with permission from Yu et al. [35]. Copyright 2017, American Chemical Society. (b) Raman spectra of WS_2 at different positions marked in the wafer-scale monolayer image. Source: Reproduced with permission from Chen et al. [36]. Copyright 2019, American Chemical Society. (c) Photograph of 1 × 1 cm SiO_2/Si covered with InSe film. Source: Reproduced with permission from Yang et al. [37]. Copyright 2017, American Chemical Society. (d) Schematic diagrams highlighting the unidirectional growth of h-BN domains and the anisotropic growth speed on a Cu surface with steps. This method obtained 100-cm^2 single-crystal hexagonal boron nitride monolayer on copper. Source: Reproduced with permission from Wang et al. [38]. Copyright 2019, Springer Nature. (e) Illustration of a chemically synthesized inverter based on $MoTe_2$. Source: Reproduced with permission from Zhang et al. [39]. Copyright 2019, Springer Nature.

electronic area, such as FET and photodetector, have been intensified in recent years. Multifunctional thermoelectric, superconducting, and magnetic devices need further investigation. For example, thermoelectric applications of 2D p–n junctions have not been thoroughly investigated yet. Giant magnetoresistance effect has been realized in CrI_3, while spin–orbit torque switching, spin Hall effect in antiferromagnets, and memory transistor based on 2D materials are rarely reported.

References

1 Novoselov, K.S., Geim, A.K., Morozov, S.V. et al. (2004). *Science* 306: 666.
2 Lev Davidovich, L. (1937). *Phys. Z. Sowjetunion* 11: 26.
3 Geim, A.K. and Novoselov, K.S. (2007). *Nat. Mater.* 6: 183.
4 Meyer, J.C., Geim, A.K., Katsnelson, M.I. et al. (2007). *Nature* 446: 60.
5 Lee, C., Wei, X.D., Kysar, J.W., and Hone, J. (2008). *Science* 321: 385.
6 Neto, A.C., Guinea, F., and Peres, N.M. (2006). *Phys. World* 19: 33.
7 Das Sarma, S., Adam, S., Hwang, E.H., and Rossi, E. (2011). *Rev. Mod. Phys.* 83: 407.
8 Neto, A.C., Guinea, F., Peres, N.M. et al. (2009). *Rev. Mod. Phys.* 81: 109.
9 Wallace, P.R. (1947). *Phys. Rev.* 71: 622.
10 McClure, J.W. (1957). *Phys. Rev.* 108: 612.
11 Slonczewski, J.C. and Weiss, P.R. (1958). *Phys. Rev.* 109: 272.
12 Novoselov, K.S., Jiang, Z., Zhang, Y. et al. (2007). *Science* 315: 1379.
13 Schedin, F., Geim, A.K., Morozov, S.V. et al. (2007). *Nat. Mater.* 6: 652.
14 Novoselov, K.S., Geim, A.K., Morozov, S. et al. (2005). *Nature* 438: 197.
15 Balandin, A.A., Ghosh, S., Bao, W. et al. (2008). *Nano Lett.* 8: 902.
16 Bolotin, K.I., Sikes, K., Jiang, Z. et al. (2008). *Solid State Commun.* 146: 351.
17 Balandin, A.A. (2011). *Nat. Mater.* 10: 569.
18 Nair, R.R., Blake, P., Grigorenko, A.N. et al. (2008). *Science* 320: 1308.
19 Zhang, Y.B., Tang, T.T., Girit, C. et al. (2009). *Nature* 459: 820.
20 Radisavljevic, B., Radenovic, A., Brivio, J. et al. (2011). *Nat. Nanotechnol.* 6: 147.
21 Wang, Q.H., Kalantar-Zadeh, K., Kis, A. et al. (2012). *Nat. Nanotechnol.* 7: 699.
22 Chhowalla, M., Shin, H.S., Eda, G. et al. (2013). *Nat. Chem.* 5: 263.
23 Wei, Z., Li, B., Xia, C. et al. (2018). *Small Methods* 2: 1800094.
24 Cui, Y., Li, B., Li, J., and Wei, Z. (2017). *Sci. China Phys. Mech. Astron.* 61: 016801.
25 Wang, X., Cui, Y., Li, T. et al. (2019). *Adv. Opt. Mater.* 7: 1801274.
26 Cui, Y., Zhou, Z., Li, T. et al. (2019). *Adv. Funct. Mater.* 29: 1900040.
27 Liu, C.-C., Feng, W., and Yao, Y. (2011). *Phys. Rev. Lett.* 107: 076802.
28 Liu, C.-C., Jiang, H., and Yao, Y. (2011). *Phys. Rev. B* 84: 195430.
29 Osada, M. and Sasaki, T. (2012). *Adv. Mater.* 24: 210.
30 Dean, C.R., Young, A.F., Meric, I. et al. (2010). *Nat. Nanotechnol.* 5: 722.
31 Ashton, M., Paul, J., Sinnott, S.B., and Hennig, R.G. (2017). *Phys. Rev. Lett.* 118: 106101.

32 Mounet, N., Gibertini, M., Schwaller, P. et al. (2018). *Nat. Nanotechnol.* 13: 246.
33 Li, H., Zhang, Q., Duan, X. et al. (2015). *J. Am. Chem. Soc.* 137: 5284.
34 Li, B., Xing, T., Zhong, M. et al. (2017). *Nat. Commun.* 8: 1958.
35 Yu, H., Liao, M., Zhao, W. et al. (2017). *ACS Nano* 11: 12001.
36 Chen, J., Shao, K., Yang, W. et al. (2019). *ACS Appl. Mater. Interfaces* 11: 19381.
37 Yang, Z., Jie, W., Mak, C.-H. et al. (2017). *ACS Nano* 11: 4225.
38 Wang, L., Xu, X., Zhang, L. et al. (2019). *Nature* 570: 91.
39 Zhang, Q., Wang, X.-F., Shen, S.-H. et al. (2019). *Nat. Electron.* 2: 164.

2

Electronic Structure of 2D Semiconducting Atomic Crystals

With recent developments in theoretical and numerical methods, as well as the growing power of supercomputers, theoretical modeling has become a more and more important tool in the research of 2D semiconductors. On the one hand, theoretical modeling can help people to understand the experimental results in a more systematic and fundamental way, thus giving insight into processes that are hidden behind the experiment. On the other hand, it can also be employed to predict new properties or design new materials, thus providing guidance to experiment and accelerate material development. For example, based on theoretical calculations, type-II band alignment between 2D semiconductors is proposed [1], and such a model is widely used to understand the optoelectronic properties in 2D heterostructures. Localized band edge states are predicted in 2D heterostructures from computation first [2] and later observed in experiments [3]. In the following three chapters, we will discuss topics related to theoretical modeling of 2D semiconductors. In this chapter, we will focus on the fundamental electronic structures of several typical 2D semiconductors and their heterostructures. In Chapter 3, we will focus on band structure modulations of 2D semiconductors. In Chapter 4, we will focus on quantum transport properties of 2D semiconductors.

2.1 Theoretical Methods for Study of 2D Semiconductors

Before discussing the detailed electronic properties of 2D semiconductors, here, we briefly introduce several methods that are commonly used in theoretical modeling.

2.1.1 Density Functional Theory

Density functional theory (DFT) reveals that the electron density of a many-electron system can solely determine the properties of this system, and the total energy is minimized by the correct ground-state electron density [4]. In DFT, the many-body problem is transformed into a single-particle Kohn–Sham equation, by attributing all the contributions of many-body effects to the exchange–correlation energy term E_{xc} [5]. DFT is exact in principle, but the actual form of E_{xc} is unknown; hence, approximate functionals for E_{xc} are usually used. It is assumed by the local density approximation (LDA) that the E_{xc} functional depends only on the value of local electronic density [5, 6]. The generalized gradient approximation (GGA) takes both the electron density and its gradient into account [7]. LDA and GGA are employed extensively to investigate the structural, mechanical, electronic, and magnetic properties of materials. Usually, they give acceptable results. However, one of the major problems of LDA and GGA is that both of them severely underestimate the bandgap resulted from the missing of derivative discontinuity of total energy at integer particle numbers. The hybrid functionals such as HSE [8] improve the total energy estimation by mixing nonlocal Hartree–Fock exchange with LDA or GGA energy. They usually give much better bandgaps than LDA or GGA does, but computational cost is significantly larger.

2.1.2 Linear Scaling Three-Dimensional Fragment (LS3DF) Method

The computational scaling of direct DFT calculations is $O(N^3)$, where N is the number of atoms in the supercell. Thus, its application to systems with over 1000 atoms is limited because of huge computational cost, and linear scaling $O(N)$ electronic structure algorithms are in high demand. The linear scaling three-dimensional fragment (LS3DF) method [9] is one of such $O(N)$ methods. It divides the system into fragments and then calculates each fragment and patches them into the original system with novel boundary cancellation techniques. The Coulomb potential, based on the global charge density, is solved on the whole system, and it thus includes all the self-consistent effects. It yields almost the same results as do direct DFT calculations and allows simulations of systems with over 10 000 atoms at a moderate computational cost.

2.1.3 GW Approximation

To date, the most suitable approach to study electronic quasiparticle excitations is the many-body perturbation theory based on the one-body Green's function [10]. In this approach, the energy-dependent and nonlocal self-energy term Σ_{xc} includes all nonclassical many-body effects. The GW method approximates the Σ_{xc} using its first-order expansion with respect to the dynamically screened Coulomb interaction W and the Green's function G [10, 11]. W and G are often calculated on the basis of the eigen states of a reference single-particle Hamiltonian, and the quasiparticle energies are calculated as a first-order correction to the single-particle eigen energies. So far, the GW method has been successfully applied to the calculation of quasiparticle band structure properties for a wide class of materials. However, it also suffers from convergence issues and unfavorable scaling of the computational cost regarding the system size.

2.1.4 Semiempirical Tight-Binding Method

Tight-binding (TB) method is primarily used for band structure calculation of a material. It uses atomic orbitals as a basis to expand the single-electron wave functions of the system. The Hamiltonian matrix elements between these atomic orbitals are treated as adjustable parameters [12] and fitted to the results of experiments or first-principles calculations, and the eigen values and eigen states are then calculated by diagonalizing the Hamiltonian matrix. Despite its simplicity, tight-binding model can give good qualitative results with much low computational cost compared to DFT calculations. A major problem of the tight-binding method is that the fitted parameters are highly system dependent; thus, the transferability is poor.

2.1.5 Nonequilibrium Green's Function Method

The nonequilibrium Green's function (NEGF) formalism is a popular method to calculate the electron or phonon transport properties of extended systems [13, 14]. In this approach, the simulated system is constructed by three parts. Two semi-infinite leads serve as the electron or heat baths, and they are connected by a central conductor region. The transmission of electron or phonon is calculated based on the Green's function for the center region, and the self-energy of the leads which describes the lead–center interaction. The main advantage of NEGF is that the quantum mechanical effects such as tunneling and diffraction are preserved, which allows a highly accurate description of nanoscale devices. However, for large devices, the NEGF method can be computationally expensive.

2.2 Electronic Structure of 2D Semiconductors

There are various types of 2D semiconductors. One class of the representatives is graphene derivative. Examples of graphene derivatives include the graphyne family members (GFM, such as graphyne and graphdiyne) and nitrogenated holey graphene (NHG) C_2N. Another class is transition metal dichalcogenides (TMDs) such as MoS_2, $MoSe_2$, WS_2, etc. In this section, we discuss the fundamental electronic structure of these two classes of 2D semiconductors. All calculations are performed based on DFT if not specified.

2.2.1 Graphyne Family Members

The structure of GFM can be viewed as a 2D network of hexagonal carbon rings (sp^2 hybridized) connected by acetylenic linkages (sp hybridized), as presented in Figure 2.1. It has a hexagonal symmetry similar to that of graphene. The length of the acetylenic linkages can be different, leading to the graphyne-n structures, in which n indicates the number of carbon triple bonds in the linkage (Figure 2.1). Graphdiyne (graphyne-2) is the first experimentally accessible member in the graphyne family [16]. Compared with graphene, the graphyne-n structures show weaker stability because inserting the acetylenic linkages into the carbon network reduces the cohesive energy [17]. Among the graphyne-n structures, graphyne is predicted to be most stable, and the cohesive energy decreases as the length of acetylenic linkage increases [17]. The coexistence of sp and sp^2 hybridization

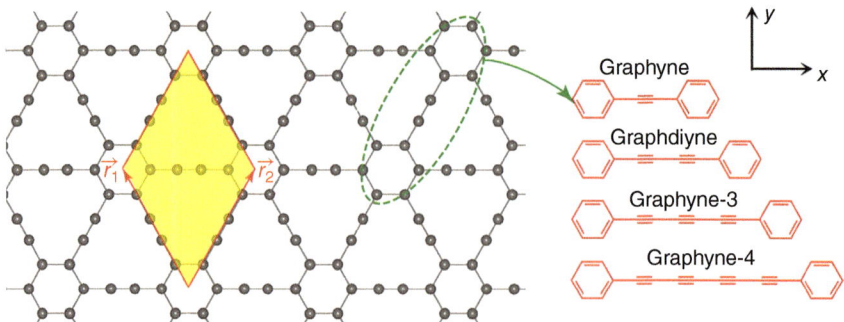

Figure 2.1 Illustration of the graphyne-*n* structures. Source: Reproduced with permission from Yue et al. [15]. Copyright 2013, American Chemical Society.

in graphyne-*n* results in different C—C bonding types. In graphyne, there are three types of C—C bonds: the C(sp^2)—C(sp^2) bond with a length of 1.43 Å, the C(sp^2)—C(sp) bond with a length of 1.40 Å, and the C(sp)—C(sp) bond with a length of 1.23 Å. In graphyne-*n* with *n* larger than 1, besides the above three types of bonds, there is another C(sp)—C(sp) bond that connects two C(sp)—C(sp) bonds, and its length is 1.33 Å [18].

The band structure of graphyne calculated using GGA–PBE is shown by the solid lines in Figure 2.2a. The bandgap is found to be 0.46 eV, located at the M point. The effective masses are also calculated using $m^* = \hbar^2/(\partial^2 E/\partial k^2)$. The results are listed in Table 2.1. It is found that the m^* values along the M–Γ direction are two times more than those along the M–K direction. However, it is well known that the bandgap of semiconductor is underestimated by LDA and GGA. In contrast, hybrid functionals such as HSE06 can reproduce the experimentally measured bandgap quite well and give excellent description of the electronic structure of semiconductor. The HSE06 calculated band structure of graphyne is presented in Figure 2.2a by open circles. A 0.96 eV bandgap is obtained, which is twice as much as the GGA–PBE value. The HSE06-calculated effective masses are also listed in Table 2.1. Compared with GGA–PBE results, HSE06 gives larger m^* along the M–K direction and slightly smaller m^* along M–Γ. Except for the bandgap, the band structures given by GGA–PBE and HSE06 exhibit similar features. Therefore, we expect that there are no obvious differences between the electronic properties calculated by these two functionals, and the following discussions are based on results obtained using GGA–PBE.

As mentioned above, there are both sp^2- and sp-hybridized C atoms in graphyne, resulting in several different bond types. The bond between two sp^2 C atoms has a $\sigma + \pi$ character. The σ bond is contributed by $p_x + p_y$ and s orbitals of C atom, and the π bond comes from the p_z orbital. Two sp C atoms are $\sigma + 2\pi$ bonded. The characters of the σ bond and one of the π bonds are the same as sp^2—sp^2 bonding, whereas the other π bond is contributed by $p_x + p_y$ orbitals, which is absent in sp^2—sp^2 bonding. The bond between C-sp^2 and C-sp is a single σ bond and has $p_x + p_y$ and s characters. In Figure 2.2b, the projection of atomic orbitals on the energy band is plotted for two C atoms, C1 and C2. C1 is on the

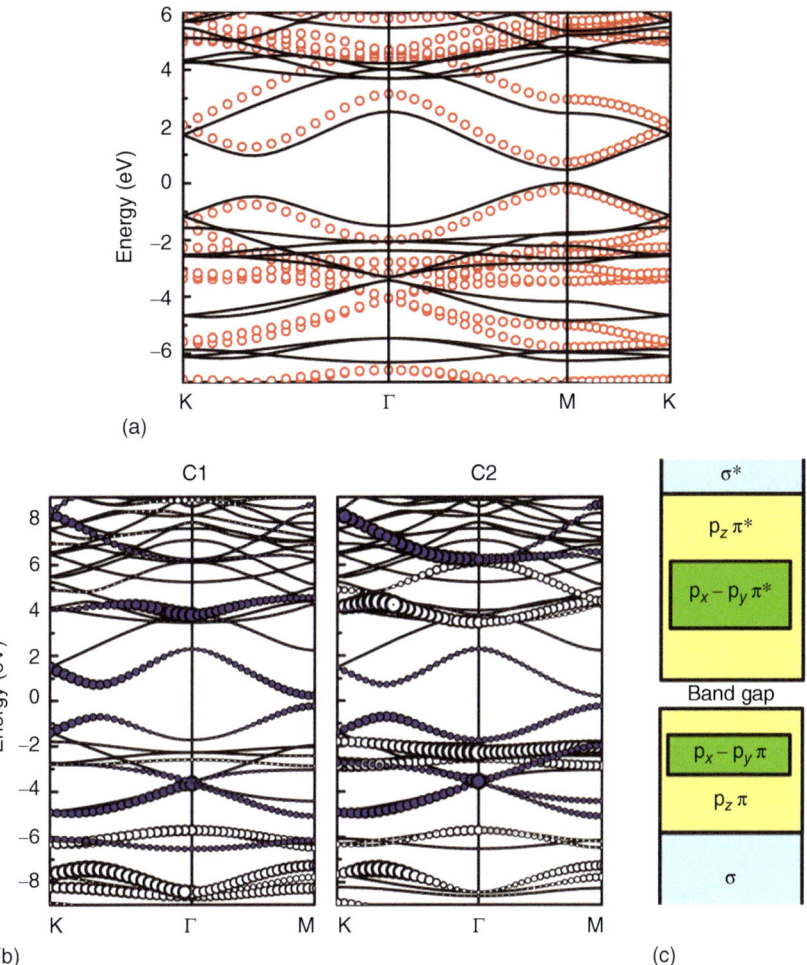

Figure 2.2 (a) Band structure of graphyne. Solid lines and open circles are GGA–PBE and HSE06 calculated results, respectively. The energy of VBM calculated by GGA–PBE is set to zero. (b) Projected band structures for two C atoms with sp^2 hybridization (C1) and sp hybridization (C2) in graphyne. Contributions from p_x and p_y orbitals are represented by open circles, and the contributions from p_z orbital are indicated by solid circles. (c) Division of energy region according to different bonding types. Source: Reproduced with permission from Kang et al. [19]. Copyright 2011, American Chemical Society.

Table 2.1 The bandgap and effective masses for graphyne.

	Bandgap (eV)	m^*_c/m_0 (Γ–M)	m^*_c/m_0 (M–K)	m^*_v/m_0 (Γ–M)	m^*_v/m_0 (M–K)
GGA–PBE	0.46	0.20	0.080	0.21	0.083
HSE06	0.96	0.18	0.11	0.19	0.12

The values obtained by both GGA–PBE and HSE06 are presented. m_0 is the electron mass.

hexagonal ring with sp^2 hybridization and C2 is on the acetylenic linkage with sp hybridization. On the basis of the projected band structure, the band structure of graphyne can be divided into several energy regions according to the bonding character, as can be seen in Figure 2.2c. The valence bands (VBs) correspond to the bonding states. In the region below 6 eV, the bands are dominated by p$_x$ and p$_y$ states of both C atoms, and they can be classified as bonding σ bands. The bonding p$_z$ π bands range from 6 eV to valence band minimum (VBM) because the p$_z$ states of both C atoms have significant contributions in this region. Besides, it is notable that several bands with p$_x$ and p$_y$ characters can be found between 3 and 2 eV. These p$_x$ and p$_y$ states mainly come from C2 and the contribution of C1 is negligible. Therefore, we can conclude that these bands are bonding p$_x$–p$_y$ π bands, which come from the bonding between two sp C atoms. The conduction bands (CBs) correspond to antibonding states. The antibonding p$_z$ π^* bands range from conduction band maximum (CBM) to $\sim$11 eV. The antibonding p$_x$–p$_y$ π^* bands appear between 4 and 6 eV. In addition, the antibonding σ^* bands are found above 9 eV. The p$_z$ π and p$_z$ π^* bands are much wider than p$_x$p$_y$ π and p$_x$–p$_y$ π^* bands, indicating that the p$_z$ π bond is much more delocalized than p$_x$–p$_y$ π bond. This can be easily understood because both sp^2 and sp C atoms have p$_z$ π bonds, and there are large overlaps between p$_z$ orbitals of adjacent C atoms. Nevertheless, p$_x$–p$_y$ π bonds exist only in sp C atoms. They are separated spatially by the hexagonal rings, leading to small overlaps.

2.2.2 Nitrogenated Holey Graphene

Nitrogenated holey graphene is another graphene derivative. It was successfully synthesized in 2015 [20]. The primitive unit cell of the NHG crystal structure, with the formula C_2N, can be constructed using two benzene rings surrounded by six nitrogen atoms. As shown in Figure 2.3a, in NHG, carbon and nitrogen atoms form an 18-atomic hexagonal unit cell with the lattice constant 8.29 Å. Similar to graphene, the strong sp^2 hybridization between carbon and nitrogen atoms leads to the formation of an atomically flat holey structure. In addition, NHG structure contains three types of bond lengths: 1.33 Å for carbon–nitrogen bonds, 1.46 Å for carbon–carbon bonds facing the holey side, and 1.43 Å for carbon–carbon bonds located in between the nitrogen linkers. It appears from the short bond lengths that, withstanding the presence of the nitrogen atoms between the benzene rings, the NHG crystal structure is quite stable. According to Bader charge analysis, the holey site of graphene is surrounded by negatively charged N atoms ($-1.1\,e$), while each C atom in the benzene rings donates about $0.55\,e$. However, strong carbon–nitrogen hybridization leads to the emergence of some additional features such as an energy bandgap. The band structure of the NHG monolayer is shown in Figure 2.3b. It has a direct bandgap located at the Γ point. Around the band edges, there are several flat bands. The contour plots of the lowest conduction band (CB) and highest valence band (VB) around the Γ point are presented in Figure 2.3c,d. Notice that the CB exhibits a much more anisotropic behavior than the VB. From Figure 2.3c,d, it can be seen that the contour lines for VB are more circle-like than those for the CB. The energy changes from Γ to K and from Γ to M are 0.08 and 0.11 eV for VB, respectively. However, for CB,

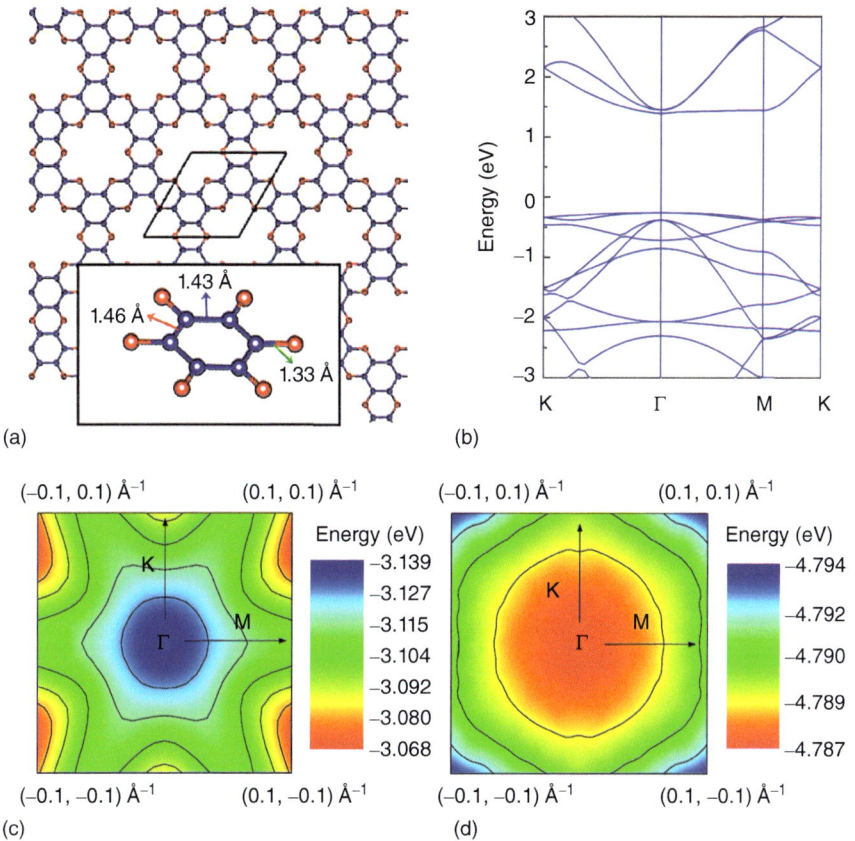

Figure 2.3 (a) Top view of the atomic structure of a single-layer holey graphene. Nitrogen and carbon atoms are shown by red and blue balls, respectively. Primitive unit cell of the crystal is delineated by black rhombus. Tilted perspective view of a N-surrounded benzene unit and bond lengths are given in the inset. (b) Band structure of NHG monolayer. (c) Contour plot of the lowest conduction band around the Γ point. (d) The same as (c) but for the highest valence band. Source: Reproduced with permission from Kang et al. [21]. Copyright 2015, American Physical Society.

the values are 0.76 and 0.06 eV. The PBE-predicted energy gap value is 1.65 eV at the Γ point, which is smaller than the experimental value of 1.96 eV, because of the well-known bandgap underestimation of PBE. The PBE-predicted effective masses are 17.25 m_0 for the hole and 1.13 m_0 for the electron along Γ–K and 22.34 m_0 for the hole and 2.77 m_0 for the electron along Γ–M.

2.2.3 Transition Metal Dichalcogenides

TMDs have the formula of MX_2, where M is a transition metal atom and X is a chalcogen atom. In these materials, the M atom layer is sandwiched between two X atom layers. For monolayers of MX_2, one of the most common phases is the 1H structure with the D_{3h} symmetry as shown in Figure 2.4a. In the present work,

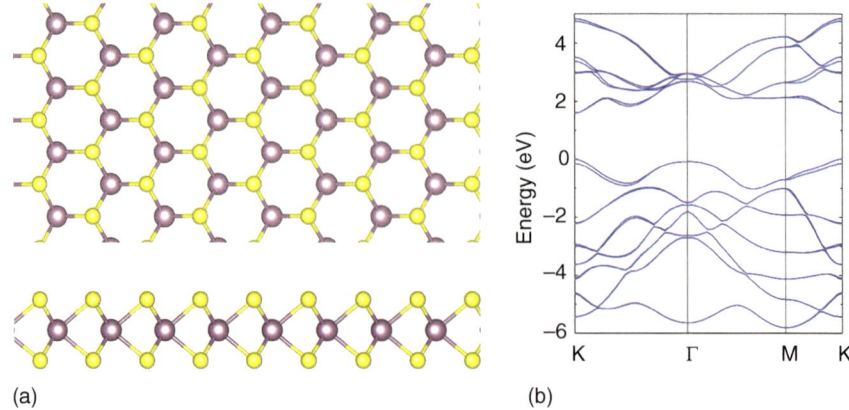

Figure 2.4 (a) Structure of 1H-MX$_2$. Yellow indicates X atoms and purple indicates M atoms. (b) Band structure of MoS$_2$ calculated by GGA–PBE with spin–orbital coupling.

Table 2.2 Calculated properties of MX$_2$ monolayers: lattice constant a, M—X bond length d, bandgap E_g, spin–orbit splitting Δ_{SO} in valence band, cohesive energy E_C per unit cell, charge transfer $\Delta\rho$ of the M atom, Poisson's ratio v, and in-plane stiffness C (a, d, E_g, and Δ_{SO} calculated by HSE06 are also shown).

	a (Å)	a^{HSE} (Å)	d (Å)	d^{HSE} (Å)	E_g (eV)	E_g^{HSE} (eV)	Δ_{SO} (eV)	Δ_{SO}^{HSE} (eV)	E_C (eV)	$\Delta\rho$ (e)	v	C (N/m)
MoS$_2$	3.18	3.16	2.41	2.40	1.59	2.02	0.15	0.20	15.31	1.09	0.25	124.24
MoSe$_2$	3.32	3.29	2.54	2.51	1.33	1.72	0.19	0.27	13.70	0.85	0.23	103.40
MoTe$_2$	3.55	3.52	2.73	2.70	0.94	1.28	0.22	0.35	12.01	0.52	0.24	78.90
WS$_2$	3.18	3.16	2.42	2.40	1.55	1.98	0.43	0.56	17.28	1.24	0.22	139.54
WSe$_2$	3.32	3.29	2.55	2.53	1.25	1.63	0.47	0.63	15.45	0.96	0.19	115.52
WTe$_2$	3.55	3.52	2.74	2.71	0.75	1.03	0.49	0.69	13.51	0.57	0.18	86.93

we focus on the most typical MX$_2$ with M = Mo and W and X = S, Se, and Te. The basic lattice parameters and physical properties of single-layer MX$_2$ are listed in Table 2.2. When X goes from S to Te, the lattice constant and bond length of MX$_2$ increase associated with the increase in ionic radius of X. The lattice constants of MoX$_2$ and WX$_2$ are very close to each other, indicating that synthesizing MoX$_2$–WX$_2$ heterostructures might allow one to engineer optical and physical properties with minimum structural defects. Finally, elastic parameters that are obtained by fitting the strain–energy relationship of MX$_2$ monolayers show that the in-plane stiffness decreases when X changes from S to Te and M changes from W to Mo. The calculated Poisson's ratios of MX$_2$ monolayer are around 0.2, and the Poisson's ratio of MoX$_2$ is found to be slightly larger than that of WX$_2$.

The band structure of MoS$_2$ is presented in Figure 2.4b. Other MX$_2$ showed similar band characters. In their bulk form, the MX$_2$ considered here exhibit an indirect bandgap [22]. However, the monolayers of MX$_2$ have direct bandgaps, with the CBM and VBM located at the K symmetry points, as well as a significant

2.2 Electronic Structure of 2D Semiconductors | 17

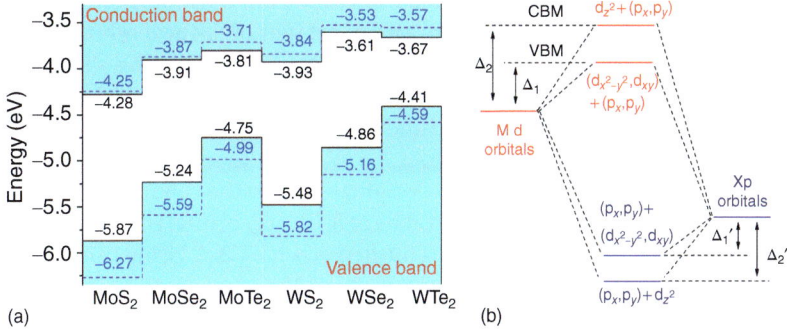

Figure 2.5 (a) Calculated band alignment for MX$_2$ monolayers. Solid lines are obtained by PBE and dashed lines are obtained by HSE06. The vacuum level is taken as zero reference. (b) Schematic of the origin of CBM and VBM in MX$_2$. Source: Reproduced with permission from Kang et al. [1]. Copyright 2013, AIP Publishing.

spin–orbit splitting in valence band, as listed in Table 2.2. Compared to the experimental values, we observe that the bandgap values are underestimated by PBE while they are slightly overestimated by HSE06. Here, we note that regardless of the choice of the method (PBE/HSE06), the observed general trends display similar behavior. Therefore, the deviation from the experimentally reported bandgap values can be omitted when comparing the band offsets between MX$_2$ monolayers, and hence, the chemical trends of band structures of MX$_2$ can be properly described by both PBE and HSE06.

The band alignment for MX$_2$ monolayers is shown in Figure 2.5a. The vacuum energy is used to align the potential between different systems. In the following, we discuss the results in Figure 2.5 based on PBE calculations, but similar results can be deduced from HSE06. Overall, the band offsets show the following trends:

1. As the atomic number of X increases (from S to Te), the energies of CBM and VBM of MX$_2$ also increase. Moreover, the conduction band offset (CBO) is smaller than the valence band offset (VBO). More specifically, the VBM of MoSe$_2$ is 0.63 eV higher than that of MoS$_2$, whereas its CBM is 0.37 eV higher than that of MoS$_2$. Here, the only exception is WTe$_2$, where its CBM is slightly lower (by 0.06 eV) than that of WSe$_2$.
2. For a common-X system, the CBM and VBM of WX$_2$ are higher than those of MoX$_2$, i.e. MoX$_2$–WX$_2$ lateral heterostructures have a type-II band alignment. For example, the VBM of WS$_2$ is 0.39 eV higher than that of MoS$_2$, and its CBM is 0.35 eV higher.

To understand the observed trends in the band offsets, we look at the physical origin of the observed CBM and VBM values. For simplifying, the following discussions are made based on the band structure calculated without spin–orbit coupling (SOC) because SOC only affects the CBM and VBM energies, while the general trends of the band offsets were found to be the same. Taking MoS$_2$ as an example, the VBM of MoS$_2$ mainly consists of the $d_{x^2-y^2}$ and d_{xy} orbitals of Mo and the p_x and p_y orbitals of S. In addition, it is found that the fifth band at the K points, which locates about 4 eV lower than the VBM, has the same character as

VBM. Therefore, the VBM of MoS_2 originates mainly from the repulsion between the $d_{x^2-y^2}$ and d_{xy} orbitals of Mo and the p_x and p_y orbitals of S. The d orbital of Mo is higher than the p orbital of S, so it is pushed up by Δ_1, forming the VBM, and the p orbital is pushed down by Δ_1', as shown in Figure 2.5b. The CBM of MoS_2 has the character of the d_{z^2} orbital of Mo and the p_x and p_y orbitals of S, while the fourth band at the K points, located about 4.5 eV lower than the VBM, has the same character. Thus, the CBM of MoS_2 originates from the repulsion between the d_{z^2} orbital of Mo and the p_x and p_y orbitals of S. The d_{z^2} orbital is pushed up by Δ_2, forming the CBM, and the p orbital is pushed down by Δ_2'. Using this model, we can understand the trends of the band offsets of MX_2.

For common-M systems, the VBO and CBO are determined by the repulsion strength Δ_1 and Δ_2 between the cation d orbitals and anion p orbitals. The magnitude of the repulsion solely depends on the overlap integral of d and p orbitals and their difference in energy. In principle, a larger overlap integral or a smaller energy difference leads to larger Δ value. For increasing X atomic number, its p orbitals become shallower. As a result, the anion with shallower p orbitals pushes the cation d orbitals upward more than the anion with deeper p orbitals. As a result, Δ_1 and Δ_2 values are larger in systems with larger atomic number of X, and the CBM and VBM would be higher. However, when X goes from S to Te, the M—X bond length also increases. This would lead to a decrease of the overlap integral between the d and p orbitals and partly counteract the increase of Δ_1 and Δ_2. It can be seen from Figure 2.2b that the repulsion between M d_{z^2} and X $p_x + p_y$ orbitals is stronger than that between M $d_{x^2-y^2} + d_{xy}$ and X $p_x + p_y$ orbitals, possibly because the former has more σ bonding character while the latter has more π bonding character. Consequently, the influence of the decrease in overlap integral on Δ_2 is larger than that on Δ_1, and the increase of Δ_2 would be smaller than that of Δ_1. The presented interpretation provides an explanation for the general trend that the CBO is smaller than the VBO for common-cation systems. Particularly, W has high d orbitals, and the M—X bond length of WTe_2 is the largest among all MX_2. As a result, from WSe_2 to WTe_2, the influence of decrease in overlap integral on Δ_2 overrides the influence of increase in the anion p orbital energy, and Δ_2 decreases. Therefore, the CBM of WTe_2 is lower than that of WSe_2.

The trends of band offset between common-X systems can be understood by the position of the d orbitals of cation. The energy of the 5d orbital of W is higher than that of the 4d orbital of Mo; therefore, the CBM and VBM of WX_2 are higher than those of MoX_2. Figure 2.5 shows that the band alignment between MoX_2 and WX_2 is type-II. In a type-II heterostructure, free electrons and holes will be spontaneously separated, which is suited for optoelectronics and solar energy conversion. Considering such possibility, we provide first calculations on MoX_2–WX_2 lateral heterostructures as shown in Figure 2.6. We first note that MoX_2 and WX_2 have very similar lattice parameters (Table 2.2), allowing one to create these heterostructures without inducing structural defects. Based on the distribution of charge densities, we observe that the electrons are confined in the MoX_2, while the holes are confined in the WX_2 side. It is, therefore, expected that spontaneous charge separation occurs when excitons diffuse to the WX_2/MoX_2 junction, a process that is needed for photovoltaics. In addition, superlattices

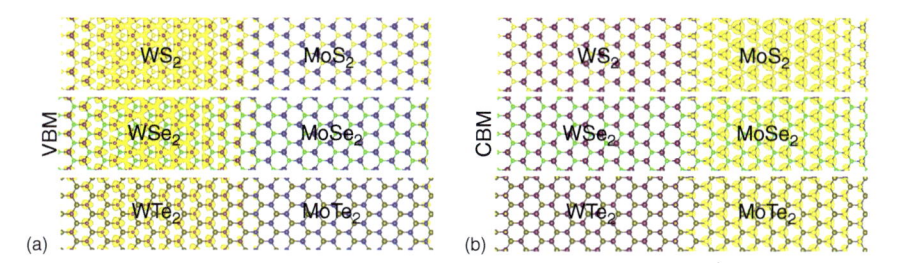

Figure 2.6 Charge densities of VBM (a) and CBM (b) states for monolayer WX_2–MoX_2 lateral heterostructures with common-X. Source: Reproduced with permission from Kang et al. [1]. Copyright 2013, AIP Publishing.

formed by interfacing alternating WX_2 and MoX_2 layers would host minibands for quantum devices, and the band offset calculated here would be needed for the design.

2.3 Prediction of Novel Properties in 2D Moiré Heterostructures

Compared to conventional semiconductor heterostructures, van der Waals 2D heterostructures are relatively easy to make (e.g. using micromechanical cleavage [23]) and have atomically sharp interfaces. One reason for this fact is the lack of requirement for lattice matching between different 2D layers because of the weak van der Waals interaction and the lack of coherent covalent bonds between the 2D layers. This, however, also raises an issue: how such lattice incommensurateness between different 2D layers will affect their electronic structures and optical properties. As a matter of fact, the 2D layers can also be rotated by almost any arbitrary angle with respect to each other [24]. This increases the variability of such 2D heterostructure design but also creates intriguing questions for their electronic and transport properties in terms of wave function localization, carrier mobility, electron–phonon coupling, and so forth. In general, lattice mismatched or rotated 2D heterostructures will show a Moiré pattern viewing from the top. Recent experiments have revealed many interesting phenomena in 2D Moiré heterostructures, such as mini Dirac cones [25, 26], localized states [27], and Moiré excitons [28]. In this section, we will use two examples, namely $MoS_2/MoSe_2$ and graphene/NHG (G/NHG), to demonstrate the effects of Moiré pattern on the electronic properties. More specifically, we will discuss the effects of lattice mismatch for $MoS_2/MoSe_2$ and the effects of rotation angle for G/NHG.

2.3.1 $MoS_2/MoSe_2$ Moiré Structure

Recently, several theoretical studies on the electronic structure of $MoS_2/MoSe_2$ heterostructures have been reported [29, 30]. In these works, the lattice mismatch between MoS_2 and $MoSe_2$ was ignored, and the monolayers were artificially compressed or stretched to form a common lateral lattice. However, as we will show later, the van der Waals binding energy between the MoS$_2$ and MoSe$_2$ layers is not strong enough to force a coherent lattice between these two layers, and hence, a nanometer-scale Moiré pattern will form. In a Moiré pattern, the stacking configurations in different regions are different, which results in spatially varying interlayer coupling strength and electrostatic potential. One intrinsic question that will be addressed in the current study is whether or not such spatial variation can cause wave function localization. If so, what is the cause of such localization: is it due to coupling variation or electrostatic potential variation? We will use first-principles calculations to study these problems. In particular, we will first analyze the effects of stacking differences by calculating different stacking patterns using DFT, and then calculate a whole 6630 atom Moiré pattern system with a linear scaling DFT method. We find that the Moiré pattern effect is strong enough to localize the hole wave function because of wave function coupling variation, while the electron wave function is weakly localized because of electrostatic potential variation. The effects of Moiré patterns suggest potential new ways to engineer the electronic structures of such systems for possible future applications.

In the current study, we will mainly focus on systems without arbitrary in-plane angular rotations. As a result, there are only two possible Moiré patterns A and B, as shown in Figure 2.7, with their in-plane hexagonal edges aligned in the same orientation and with a 4.4% lattice mismatch. As a matter of fact, one pattern can

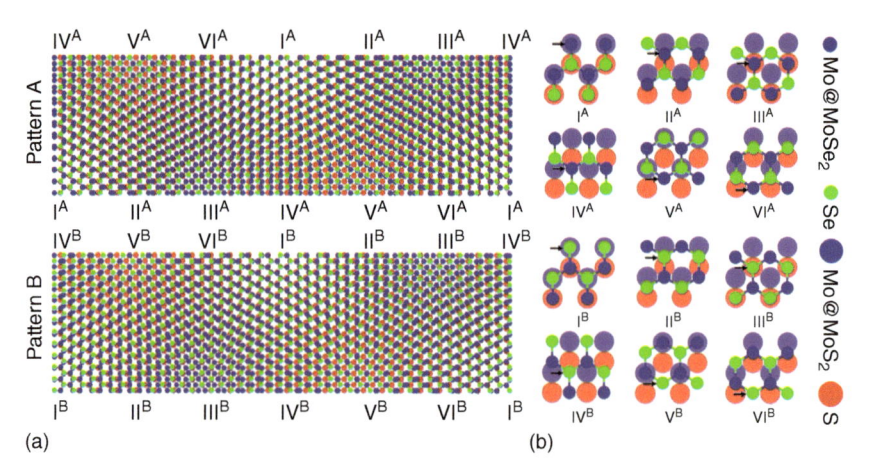

Figure 2.7 (a) Moiré patterns A and B with 4.4% lattice mismatch corresponding to 24 × 24 MoS$_2$ and 23 × 23 MoSe$_2$ supercells along the primary cell lattice vectors. We have taken a rectangular supercell (which contains two 24 × 24/23 × 23 supercells) for linear scaling DFT calculation. Only half (in (n,0) direction) of the rectangular periodic supercell is shown above to denote different stacking regions. (b) The 12 different stacking configurations considered (with MoSe$_2$ on top of MoS$_2$). Between consecutive configurations within I^A–VI^A or I^B–VI^B, the MoSe$_2$ layer is translated along the (n,n) direction in steps of $(3a)^{1/2}/6$, where a is the lattice constant, which is fixed at the average of the values for the MoS$_2$ and MoSe$_2$ monolayers. The arrows indicate the same atom in the MoSe$_2$ layer in I^A–VI^A or I^B–VI^B. Configurations I^A–VI^A cover the A pattern and configurations I^B–VI^B cover the B pattern. Source: Reproduced with permission from Kang et al. [2]. Copyright 2013, American Chemical Society.

also be considered as a result of the other by a rotation of 60° of the top MoSe$_2$ layer. In each Moiré pattern, the local stacking geometry can be classified into six different types as I^A–VI^A or I^B–VI^B, respectively, as shown in Figure 2.7. To study the effect of each stacking pattern, we have constructed small periodic systems using the average lattice constant 3.25 Å of MoS$_2$ and MoSe$_2$ to make them lattice matched. The calculations for these small systems (Figure 2.7b) are performed using DFT with SOC and the effect of van der Waals interaction [31].

In the I^A configuration, the Mo and Se atoms in the MoSe$_2$ layer are on top of the Mo and S atoms in the MoS$_2$ layer, respectively, while in the I^B configuration, the Mo and Se atoms in the MoSe$_2$ layer are on top of the S and Mo atoms in the MoS$_2$ layer, respectively. Starting from the I^A (I^B) configuration, the other configurations (before relaxation) II^A–VI^A (II^B–VI^B) can be obtained by translating the top MoSe$_2$ layer along the (n,n) (hexagonal primary cell direction index) direction (downward direction in Figure 2.7) in a step of $(3a)^{1/2}/6$ for each translation (where a is the lateral lattice constant). The adsorption energies per formula unit (each MoSe$_2$/MoS$_2$) and interlayer Se and S atom vertical height differences (after atomic relaxation) for different stacking configurations are listed in Table 2.3. The most stable configuration is III^A, with an adsorption energy of 195 meV, followed by I^B (194 meV) and V^A (191 meV). In these configurations, the Se atoms are on top of the hollow sites of the MoS$_2$ layer. Configurations with Se atoms on top of the S atoms, namely, I^A and III^B, have the lowest adsorption energies. The Se and S atom interlayer vertical height difference in different configurations varies from 3.08 to 3.77 Å and has a reverse linear relationship with respect to the adsorption energy.

Although in many cases Moiré patterns can be formed based on synthesis kinetics (e.g. the mechanical exfoliation method), it might still be interesting to discuss its formation based on total energy consideration, as done in Ref. [32], especially for synthesis methods mainly based on thermodynamics [25, 33]. In the MoS$_2$/MoSe$_2$ bilayer, if a lattice-matched structure is formed, the system can have the maximum adsorption energy using stacking pattern III^A; on the other hand, it will cost elastic energy because of lattice deformation. For a Moiré pattern heterostructure, the adsorption energy will be the average of the different stacking patterns (from I^A to VI^A or from I^B to VI^B), which turns out to be 160 meV per formula unit for Moiré pattern A and 157 meV for Moiré pattern B. These are 35 and 37 meV less than the adsorption energy of configuration III^A and I^B, respectively. On the other hand, the strain energy to stretch MoS$_2$ to the average lattice constant (from 3.18 to 3.25 Å) is 37 meV per formula unit according to DFT calculations and to compress MoSe$_2$ (from 3.32 to 3.25 Å) is

Table 2.3 Layer distance d and adsorption energy E_{ad} per formula unit (each MoSe$_2$/MoS$_2$) for different configurations.

	I^A	II^A	III^A	IV^A	V^A	VI^A	I^B	II^B	III^B	IV^B	V^B	VI^B
d (Å)	3.72	3.48	3.08	3.22	3.12	3.45	3.12	3.43	3.77	3.48	3.16	3.25
E_{ad} (meV)	118	141	195	174	191	141	194	143	120	140	177	170

42 meV per formula unit. Hence, the total strain energy (79 meV per formula unit) is much larger than the gain in the adsorption energy. As a result, a lattice-mismatched Moiré pattern structure will be formed.

The formation of Moiré patterns can lead to interesting electronic structures in $MoS_2/MoSe_2$ bilayer. To gain a better understanding, we first look at the electronic structures of the different stacking patterns listed in Figure 2.7. In the following, we will focus on Moiré pattern A, expecting similar qualitative results for Moiré pattern B. The overall band structures of all the stacking configurations are similar, and those of I^A and III^A are illustrated in Figure 2.8. Because of the absence of inversion symmetry, the band structure shows significant spin–orbit splitting that does not appear in pure MoS_2 or $MoSe_2$ bilayers [34, 35]. The current results have a direct bandgap at the K point, much as in the single-layer structure. To show the interlayer electron coupling, the projected weights of the MoS_2 and $MoSe_2$ layers to the electron wave function at a given K point and band state are denoted by colors in Figure 2.8a. It can be seen that the interlayer coupling is negligible to the band edge states around the K point. This happens because, in monolayer MoS_2 and $MoSe_2$, the band edge states at the K points are mainly localized at the central Mo layer. The highest valence band state at the K point (V1) is contributed only by the $MoSe_2$ layer, while the lowest conduction band state (C1) is contributed only by the MoS_2 layer. This observation is consistent with a type II band alignment between MoS_2 and $MoSe_2$ as discussed in Section 2.2.3 and shown schematically in Figure 2.8b, obtained by aligning the band energies of the single-layer MoS_2 and $MoSe_2$ with respect to the vacuum level. The charge density plot of the V1 and C1 states are shown in Figure 2.8a. It is also notable that the highest valence band at the Γ point (V2) has significant contributions from both the MoS_2 and $MoSe_2$ layers (green color), which is consistent with its charge density plot in Figure 2.2a. This indicates that there is a strong interlayer electron coupling effect for the V2 state. As we can see from

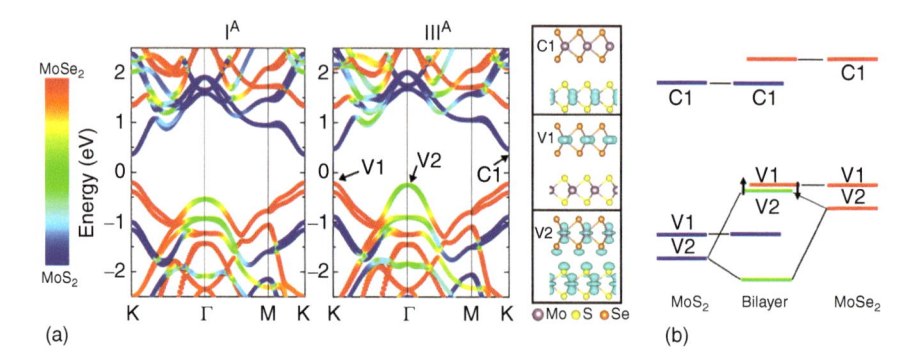

Figure 2.8 (a) Band structures of the I^A and III^A bilayers. Blue and red denote the contributions from MoS_2 and $MoSe_2$ layers, respectively. The partial charge densities of C1, V1, and V2 states of III^A are also shown. (b) Schematic of the band alignment between MoS_2 and $MoSe_2$ and the coupling effect for the V2 state in III^A. The arrows denote the eigen energy shift of V1 and V2 states when MoS_2 and $MoSe_2$ layers relax to their natural lattice constants from the average lattice constant. Source: Reproduced with permission from Kang et al. [2]. Copyright 2013, American Chemical Society.

Figure 2.8a, at different stackings (e.g. I^A and III^A), such a coupling effect can be rather different, and the coupling strength is largest when the interlayer Se and S vertical height distance is the smallest (stacking III^A). The coupling tends to push the V2 energy up as schematically shown in Figure 2.8b. For example, in a single-layer MoSe$_2$ at the average lattice constant (3.25 Å), the energy difference between V1 and V2 is 0.58 eV, while in the heterostructure in III^A, the energy difference is only 0.03 eV.

One must be cautious because the V1 and V2 levels in III^A are very close, and thus, their order could potentially be changed by higher-level computational methods (e.g. GW). Nevertheless, it was found under DFT calculation [36] that if the bilayers consisted of the same materials (e.g. two MoS$_2$ layers or two MoSe$_2$ layers), then V2 is above V1 in agreement with experiments, which provides a test for the reliability of the theoretical calculation. In our case, one can further estimate the effect when the MoSe$_2$ and MoS$_2$ layers relax to their natural lattice constants in a Moiré pattern. This can be done by calculating the absolute lateral lattice deformation potentials $E^{DP} = dE(\varepsilon)/d\varepsilon$, where $\varepsilon = (a-a_0)/a_0$ (a_0 is the equilibrium lattice constant) and $E(\varepsilon)$ is the energy of the band edge state relative to the vacuum level at a certain ε. According to our calculation, the E^{DP} of V1 and V2 states in MoSe$_2$ are −5.0 and 3.3 eV, respectively. Therefore, when relaxed from the average lattice constant in stacking III^A, the V1 state energy will shift downward by 0.11 eV, while the V2 will shift upward by 0.07 eV or slightly smaller value because of its MoS$_2$ component. Thus, according to this analysis, the V2 state in the III^A region will become the top of valence band of the whole system.

To verify the above prediction from the analysis of the stacking patterns, we have calculated a whole Moiré pattern system using the LS3DF method [9]. We have taken a rectangular supercell of Moiré pattern A (which contains two 24 × 24 MoS$_2$ and 23 × 23 MoSe$_2$ supercells with 6630 atoms in total) for LS3DF calculations. The interlayer distances in different regions of the structure are determined by an interpolation scheme from the values of the I^A–VI^A stacking patterns calculated above. The LS3DF method yields the total charge density and total potential. The folded spectrum method (FSM) [37] is used to calculate the VBM and CBM states of the whole system based on LS3DF-obtained total potential (hence, the single-particle Hamiltonian), and spin–orbit coupling is included. The calculated VBM and VBM-1 (degenerated) states are shown in Figure 2.9. It can be seen that the VBM is indeed strongly localized in the III^A region as predicted from the stacking pattern analysis. In addition, the VBM states distribute in both MoS$_2$ and MoSe$_2$ layers, much like the V2 state shown in Figure 2.8. This confirms that the VBM state of the whole system is V2 from the III^A region, as predicted above.

To further illustrate this point, we plot the distribution of the energy of the V2 state in Figure 2.10a, based on the calculations of I^A–VI^A. We align the energies of the V2 states (Figure 2.10a) in I^A–VI^A of Figure 2.7b by taking the average of the vacuum levels at the two sides of the bilayer and using that as a common energy reference. Then, we do an interpolation to obtain the landscape of the energy of the V2 state in the Moiré pattern, as shown in Figure 2.10a. Indeed, the energy of V2 is highest in III^A, where the VBM of the whole system localizes.

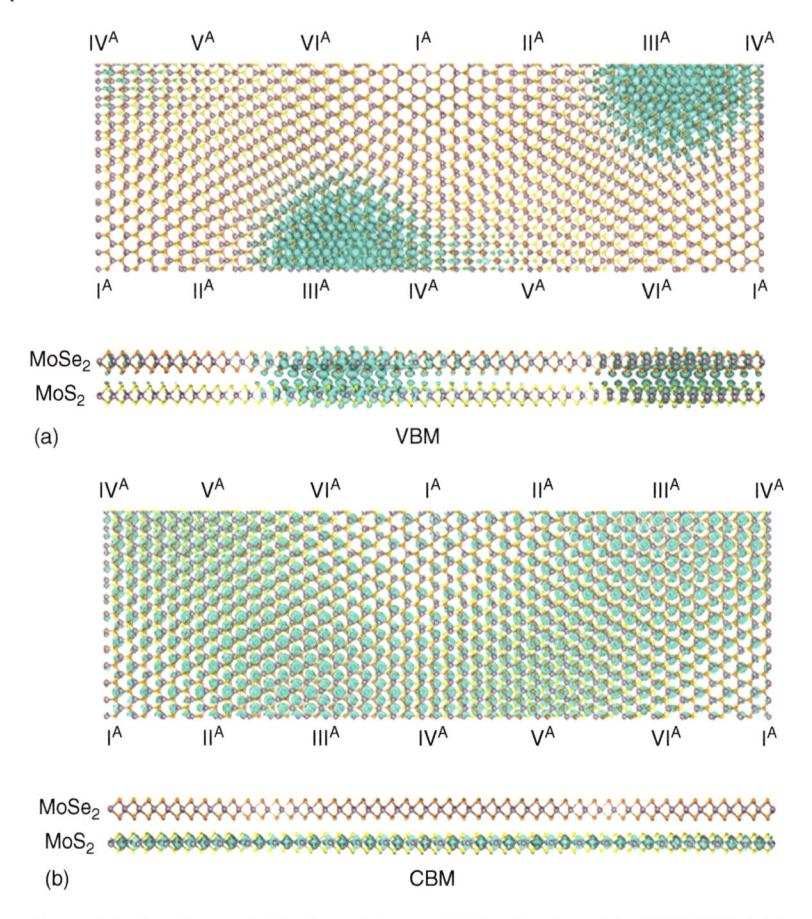

Figure 2.9 Top view and side view of the spatial distribution of the VBM, VBM-1 (a), and CBM (b) states for the Moiré structure A shown in Figure 2.7. Source: Reproduced with permission from Kang et al. [2]. Copyright 2013, American Chemical Society.

It is also interesting to note that localization of Dirac electrons is observed in rotated bilayer graphene with small twist angles (∼3°) [27, 38, 39]. Rotation in bilayer graphene also forms Moiré patterns, and small twist angles can lead to large pattern domains, much like the Moiré pattern caused by small lattice mismatches. However, the origin of the wave function localization for graphene bilayer and the $MoS_2/MoSe_2$ bilayer here differs. In graphene, the localization is related to the overlap of the displaced Dirac cones, which results in van Hove singularities [39]. In contrast, the localization in the $MoS_2/MoSe_2$ bilayer is caused by different coupling strengths at different regions of the pattern, and there is no Dirac cone in the electronic structure.

The situation of the CBM state is quite different, and its charge density is shown in Figure 2.9. It is distributed in the MoS_2 layer only, and it agrees with the C1 state at the K point as shown in Figure 2.8. The CBM wave function is only weakly localized. As there is no interlayer coupling (the wave function is localized within one layer), the weak localization must be caused by electrostatic

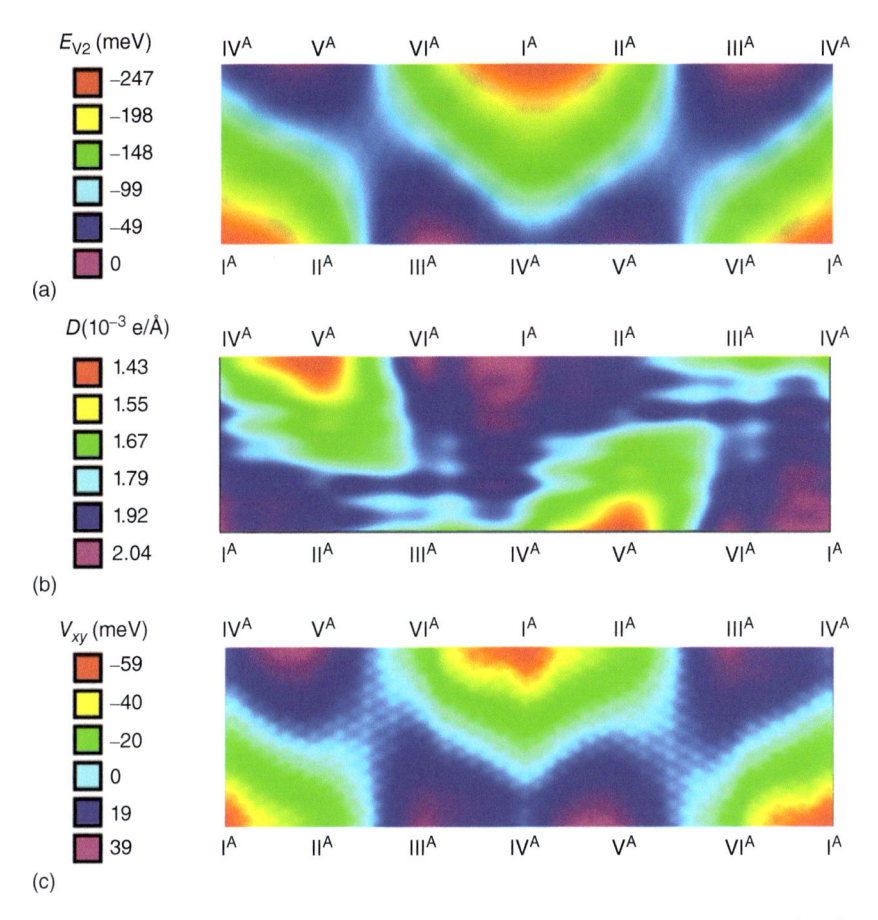

2.3 Prediction of Novel Properties in 2D Moiré Heterostructures | 25

Figure 2.10 Distribution of the energy of the V2 state based on the calculations of I^A–VI^A (a), the planar-averaged dipole moment D (b), and z-integrated local potential V_{xy} (c) in Moiré pattern A. Source: Reproduced with permission from Kang et al. [2]. Copyright 2013, American Chemical Society.

potential variations. We first calculate the charge difference $\Delta \rho(x, y, z)$ between the Moiré structure and isolated MoS_2 and $MoSe_2$ layers, with their atomic positions taken from the bilayer positions. Then, the x–y planar-averaged electron charge difference $\Delta \rho_{ave}(z)$ within each unit cell of the MoS_2 layer is calculated. The vertical direction dipole moment due to the charge transfer is then calculated as $D = -\int \Delta \rho_{ave}(z) z dz$, with the zero of z defined at the center of the bilayer. This stacking-dependent dipole moment is shown in Figure 2.10b. A positive dipole moment will mean an electron charge flow from the $MoSe_2$ layer to the MoS_2 layer, hence a higher MoS_2 CBM position. As a result, the CBM will be localized at the smallest D regions in the Moiré pattern, which agrees with the CBM localization shown in Figure 2.9. Another possible charge transfer is in the lateral direction. However, because of the highly atomic oscillation behavior, the possible lateral charge transfer in $\Delta \rho(x, y, z)$ is hard to identify. Instead, we have calculated the lateral potential averaged in the z direction:

$V_{xy} = \int [V(x, y, z)/L_z]dz$, with V being the local potential of the system and L_z the length of the periodic box in the z direction. This V_{xy} potential is then smoothed by a Gaussian convolution in the x, y directions to remove the atomic oscillations. The resulting potential is shown in Figure 2.10c. Note that any z-direction dipole moment effect will be canceled out during the z-direction average. Thus, V_{xy} should be caused by the lateral charge transfer. According to this, the CBM should be localized in the negative V_{xy} region, that is, at I^A. This is just the opposite of the z direction dipole moment effect. Judging from Figure 2.9b, at the end, it is the z direction dipole moment that determines the CBM localization.

2.3.2 Graphene/Nitrogenated Holey Graphene Moiré Structure

Besides lattice mismatch, rotation angle can also affect the properties in 2D semiconductor heterostructures. In this section, we will take G/NHG heterostructure as an example to show the effect of rotation angle.

2.3.2.1 Atomic Structure: Ordered Stacking Versus Moiré Pattern

First, we briefly describe the construction of G/NHG heterostructure. The two in-plane primitive lattice vectors of graphene are chosen as $\vec{a}_1 = (1, 0)a_0^G$ and $\vec{b}_1 = \left(\frac{1}{2}, \frac{\sqrt{3}}{2}\right)a_0^G$. For NHG, the primitive lattice vectors are $\vec{a}_2 = \left(\frac{\sqrt{3}}{2}, -\frac{1}{2}\right)a_0^N$ and $\vec{b}_2 = \left(\frac{\sqrt{3}}{2}, \frac{1}{2}\right)a_0^N$. Here, $a_0^G = 2.46$ Å and $a_0^N = 8.29$ Å are the DFT-optimized lattice constants of graphene and NHG. When integers (m,n) and (p,q) meet the condition $|m\vec{a}_1 + n\vec{b}_1| \approx |p\vec{a}_2 + q\vec{b}_2|$, a G/NHG heterostructure can be constructed, with its two lattice vectors being $m\vec{a}_1 + n\vec{b}_1 \approx p\vec{a}_2 + q\vec{b}_2$ and $-n\vec{a}_1 + (m+n)\vec{b}_1 \approx -p\vec{a}_2 + (p+q)\vec{b}_2$. We denote it as (m,n)–(p,q) stacking. Because the lattice of NHG can also be viewed as a honeycomb structure with some missing hexagonal rings, it also has zigzag and armchair directions like graphene. Therefore, one can define the relative twist angle θ in a (m,n)–(p,q)-G/NHG as the angle between the zigzag (or armchair) directions of the NHG layer and the graphene layer. Based on the above-defined lattice vectors, the cosine of θ can be calculated as:

$$\cos\theta = \frac{\sqrt{3}(mp + mq + np)}{2\sqrt{(m^2 + mn + n^2)(p^2 + pq + q^2)}}$$

As NHG has a holey honeycomb lattice, one may expect that, similar to bilayer graphene, the G/NHG heterostructure can exhibit different favorable stacking configurations such as the AA or AB stacking shown in Figure 2.11. Regarding the notation described above, AA stacking corresponds to (2,−2)–(1,0)-G/NHG with $\theta = 0°$ and AB stacking corresponds to (2,2)–(1,0)-G/NHG with $\theta = 60°$. In these cases, the lattice mismatch between the graphene layer and the NHG layer is about 3%. Formation of a lattice-matched structure imposes strain to the monolayers, causing extra strain energy. Except for the ordered stacking, another possibility is the formation of a Moiré pattern, in which the lattice mismatch between the two constitute layers is maintained, as observed in graphene/boron

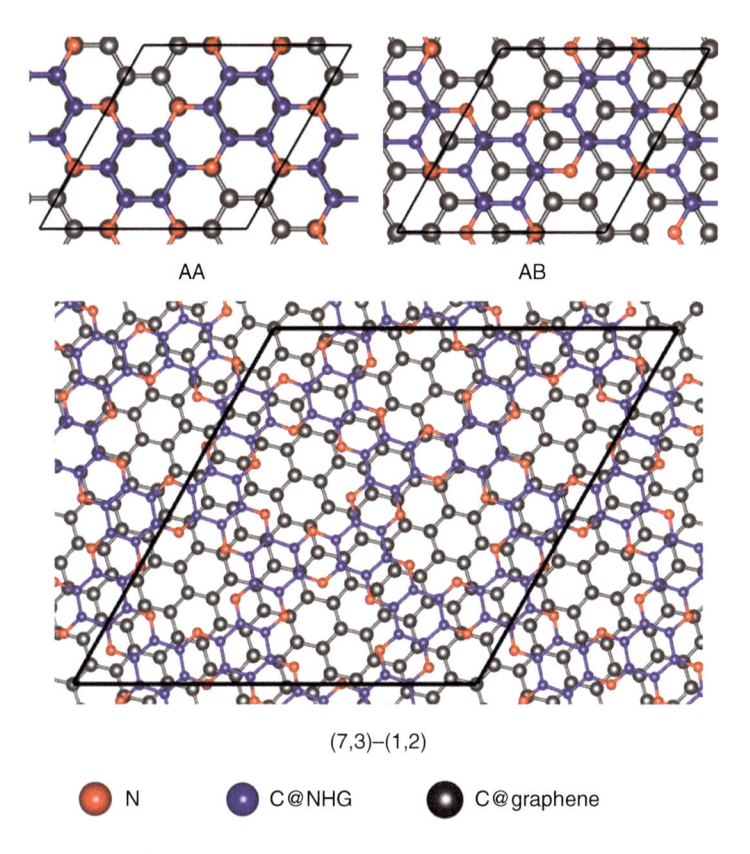

Figure 2.11 Illustration of the G/NHG heterostructures with AA, AB, and (7,3)–(1,2) stacking. The boxes defined by the solid lines indicate the supercell. Source: Reproduced with permission from Kang et al. [21]. Copyright 2015, American Physical Society.

nitride (BN) heterostructures [25, 26, 40]. In Figure 2.11, we show an example of such a Moiré pattern, namely the (7,3)–(1,2)-G/NHG. Here, the lattice mismatch is less than 0.5%, and the strain energy is negligible. The relative twist angle for this structure is 6.10°. In all these stacking configurations, graphene and NHG crystals maintain their planar crystal structure, and we performed DFT calculations to study their structural and electronic properties. The interlayer distance is calculated to be 3.36, 3.15, and 3.25 Å for AA, AB, and (7,3)–(1,2) stacking, respectively. Whether the ordered stacking or the Moiré pattern is more stable depends on the competition between the interlayer binding energy and the strain energy [32].

The interlayer binding energy is defined by $E_b = (E_{tot} - E_G - E_{NHG})/S$, where E_{tot} and S are the total energy and the area of the heterostructure supercell, and E_G and E_{NHG} are the energies of isolated graphene and NHG layers with the same lattice constant as the heterostructure, respectively. The strain energy is calculated by $E_s = (E_G + E_{NHG} - E_0^G - E_0^{NHG})/S$, where E_0^G and E_0^{NHG} are the energies of graphene and NHG layers at their equilibrium state, respectively. For different structures, the one with the lowest $E_b + E_s$ is the most stable. For AA, AB, and (7,3)–(1,2) stacking, the interlayer binding energies are −13.41, −16.52, and −15.27 meV/Å², respectively. On the other hand, the strain energy is 5.68 meV/Å² for AA and AB stacking but only about 0.1 meV/Å² for the (7,3)–(1,2) stacking. Overall, the $E_b + E_s$ for (7,3)–(1,2) stacking is the lowest. Although AB stacking has the largest gain in the interlayer binding energy, it is compensated by the high strain energy. Therefore, in G/NHG heterostructures, a lattice mismatch will be maintained, and the formation of Moiré patterns, such as the (7,3)–(1,2) structure, is favored over AA and AB stacking.

As the formation of Moiré patterns is more favorable, we take the (7,3)–(1,2) structure as an example to discuss the band structure of G/NHG. Although AA and AB G/NHG are not the ground state (they can possibly be metastable configurations), to see the effect of ordered stacking, we first look at their band structures, as shown in Figure 2.12. Because of the Brillouin zone folding, the K point of graphene is folded into the Γ point in these structures. Both structures show a bandgap opening at the Γ point. The bandgap is 0.26 eV for AA stacking and 0.19 eV for AB stacking. However, the situation for the (7,3)–(1,2) structure is very different, as can be seen in Figure 2.13a. A linear band dispersion is observed around the K point, with only a negligible bandgap opening of 0.3 meV. The bands around the K point are also found to be isotropic. Therefore, the Dirac-fermion-like behavior of carriers in graphene is preserved in G/NHG Moiré structures.

The qualitative differences in the ordered stacking and the Moiré structures can be understood as follows. The AA and AB structures are both commensurate states, in which the graphene lattice follows the periodic potential U created by the NHG layer. This potential results in the scattering between states at K and K−$\vec{G}$ in the graphene layer, where $\vec{G}$ is the reciprocal lattice vector of U and the strength depends on the $\vec{G}$ Fourier components of U [41]. In the commensurate state, $\vec{G}$ is also the reciprocal lattice vector of the primitive graphene cell.

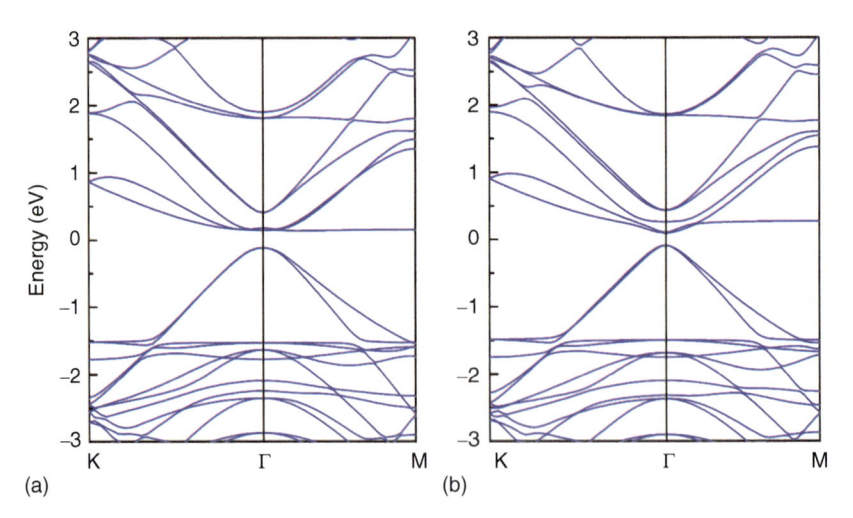

Figure 2.12 Band structure of the G/NHG heterostructures with (a) AA and (b) AB stacking. The Fermi level is set to zero. Source: Reproduced with permission from Kang et al. [21]. Copyright 2015, American Physical Society.

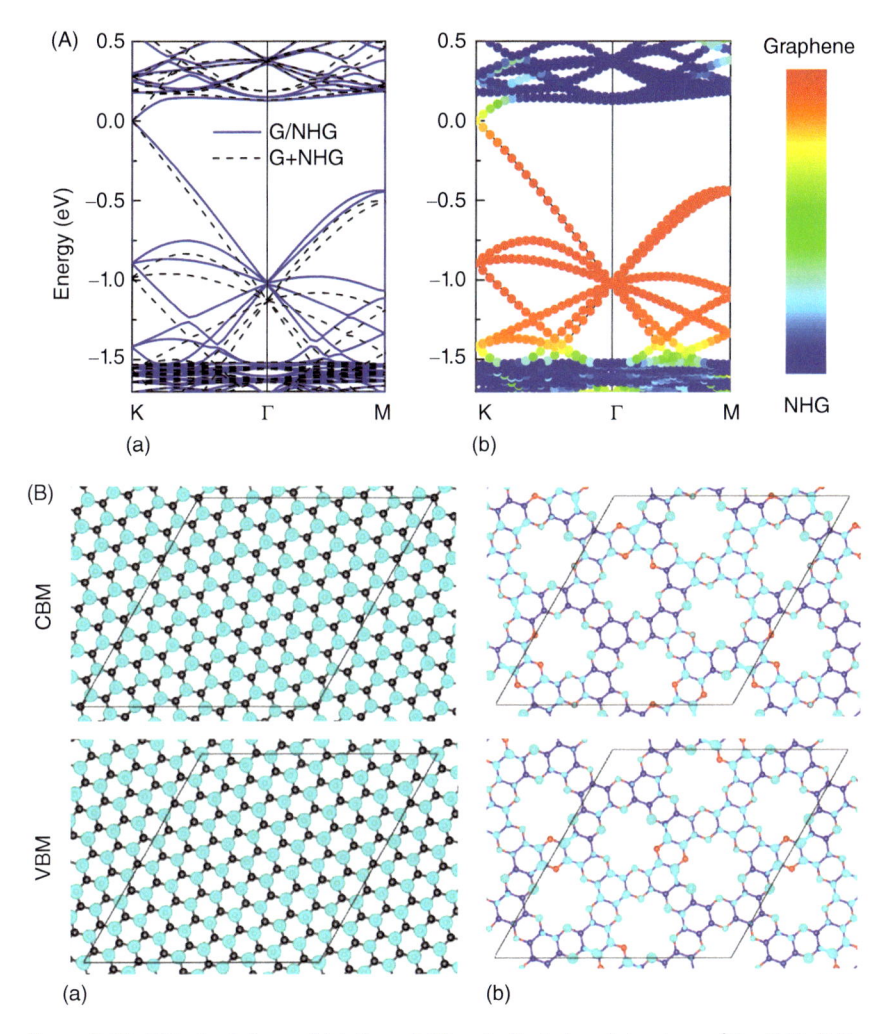

Figure 2.13 (A) In the left panel (a), the solid lines indicate band structure of the (7,3)–(1,2) structure (G/NHG), and the dashed lines are the overlap of the bands of isolated graphene and NHG layers (G + NHG). In the right panel (b), the layer projection is given. Blue and red denote the contribution from the NHG and graphene layers, respectively. (B) The partial charge density of the VBM and CBM states at the K point. Left (a) and right (b) panels show the distribution in the graphene and NHG layers, respectively. The isosurface corresponds to 0.00035 e/Å^3. Source: Reproduced with permission from Kang et al. [21]. Copyright 2015, American Physical Society.

This causes scattering between states at the K point and the K–$\vec{G}$ point, namely between the two Dirac cones. Therefore, there will be coupling between the VBM state at one cone and the CBM at the other [41], and consequently, the CBM and the VBM are no longer degenerate, leading to the opening of a bandgap. In the Moiré structures, the lattice periodicity of graphene and NHG (thus the potential U) are incommensurate. In this case, $\vec{G}$ is not a reciprocal lattice vector of graphene, and scattering between different Dirac cones do not occur. Hence, the

VBM and CBM states at the K point remain degenerate, and there is no bandgap opening. Note that in practical calculations of Moiré structures, because of the periodic boundary condition used, graphene and NHG share the same supercell and some $\vec{G}$ may also be the reciprocal lattice vector of graphene. However, the Fourier components of U at these $\vec{G}$ are rather small because they correspond to high-frequency contributions. Therefore, the scattering between Dirac cones, and the resulting bandgap opening, will be negligible. A similar effect is also predicted for graphene/BN heterostructures. In graphene/BN with a commensurate lattice (such as AA or AB stacking), there is a significant bandgap opening [42, 43], whereas in graphene/BN heterostructures with a Moiré pattern [32, 44], the bandgap is zero.

Although the van der Waals interaction between graphene and NHG is not strong, it can have a significant influence on the band structure of G/NHG. The interaction introduces a dipole moment that leads to an energy shift between the two layers. According to our calculations, the work function of an isolated graphene layer is 4.24 eV, whereas the electron affinity of an isolated NHG layer is 4.41 eV. In other words, the CBM of NHG is 0.17 eV lower than the Dirac point of graphene. If there is no interaction between graphene and NHG, when they are stacked together, the Dirac point of graphene will be buried inside the conduction bands of NHG. However, because of the interlayer coupling, there is charge transfer from graphene to NHG, resulting in a dipole moment. This dipole moment moves up the energy of the NHG layer with respect to the graphene layer. Consequently, in G/NHG, the Dirac point of the graphene layer appears inside the bandgap of the NHG layer, as shown in Figure 2.13a. In Figure 2.13a, we also show the overlap of the bands of isolated graphene and NHG layers (G + NHG), and the energy shift caused by the dipole moment is included. Comparing the bands of G/NHG and G + NHG, it is found that the general shapes of the bands are similar, but the slope of the linear bands around the K point is reduced in the G/NHG. This suggests a renormalization of Fermi velocity caused by the interlayer coupling, which we will discuss in detail later.

To further show the interlayer coupling effect, the projected weights of the graphene and NHG layers to the electron wave function at a given K point and band state are denoted by different colors in Figure 2.13a. The flat bands of the NHG layer appear at $E > 0.1$ eV and $E < -1.5$ eV. Within the range -1.5 eV $< E < 0.1$ eV, the bands originate mostly from the graphene layer. However, around the Fermi level, strong interlayer coupling is observed. It can be seen that the states close to the K point around 0 eV are not only from the graphene layer but also from the NHG layer. Hence, the Dirac states are distributed not only in the graphene layer but also in the NHG layer. In Figure 2.13b, the charge density of the VBM and CBM states at the K point are shown. These states have a p_z character. In the graphene layer, the VBM and CBM are distributed on different sublattices. Both states also have significant components in the NHG layer. Monolayer NHG is a semiconductor with a considerable bandgap. As a consequence of the formation of a Moiré heterostructure with graphene, the interlayer coupling brings Dirac-fermion-like carriers to the semiconducting NHG layer.

2.3.2.2 Renormalized Fermi Velocity

Previous studies showed that when graphene is subject to a periodic potential, its Fermi velocity decreases [41, 45]. In graphene heterostructures with Moiré patterns, a periodic potential that follows the Moiré pattern is created; thus, the Fermi velocity of graphene is renormalized, as was shown in the case of twisted bilayer graphene [27, 39]. In G/NHG heterostructures, because of the appearance of a Moiré pattern, such renormalization is also expected. Figure 2.14a shows that the slope of the linear bands around the K points in the (7,3)–(1,2) structure is smaller than that in a pristine graphene monolayer. In other words, the Fermi velocity v_F of the (7,3)–(1,2) structure is reduced. Moreover, in graphene, the v_F is the same for electrons and holes, and the calculated value is 8.46×10^5 m/s. However, the (7,3)–(1,2) structure exhibits electron–hole asymmetry. The calculated v_F is 6.24×10^5 m/s for holes and 6.02×10^5 m/s for electrons. Defining an asymmetry factor $\varphi = 2(v_F(h) - v_F(e))/(v_F(h) + v_F(e))$, the asymmetry in the (7,3)–(1,2) structure is 3.6%. In Moiré pattern heterostructures, the relative angle θ between the layers can be arbitrary in principle. Different θ may result in different properties. For example, in bilayer graphene, the magnitude of the Fermi velocity renormalization depends on the relative twist angle [27, 39]. A smaller angle leads to a larger Moiré pattern and a stronger renormalization.

To investigate Moiré-patterned G/NHG heterostructures with different θ, we have performed tight-binding (TB) calculations. We limited studies to the p_z orbital for both C and N atoms because we are mainly interested in the states around the Fermi level, which are all p_z orbitals. The Hamiltonian has the form [46]:

$$H = \sum_i \varepsilon_i |i\rangle\langle i| + \sum_{i \neq j} t_{ij} |i\rangle\langle j|$$

where $|i\rangle$ is the p_z orbital of the atom located at r_i, ε_i is the onsite energy, and t_{ij} is the coupling parameter. In monolayer graphene, only the ppπ interaction

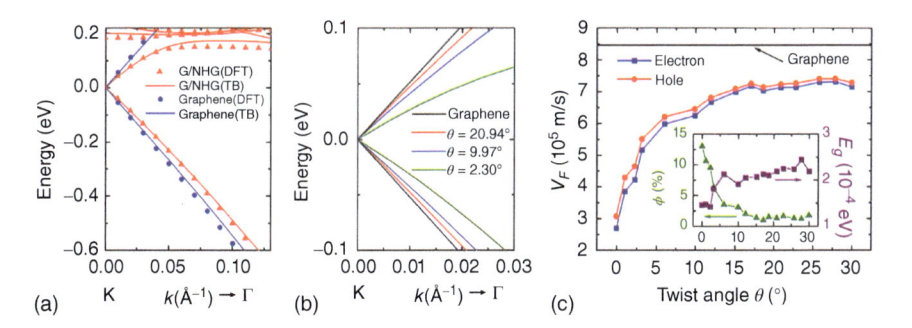

Figure 2.14 (a) Band dispersion of (7,3)–(1,2) G/NHG and graphene near the Fermi level calculated by DFT and TB. (b) Band dispersion of G/NHG with different twist angles θ around K. $\theta = 20.94°$, $9.97°$, and $2.30°$ correspond to (20,6)–(5,3), (8,15)–(0,6), and (20,1)–(3,4) stacking, respectively. (c) The calculated Fermi velocity v_F as a function of θ. Solid horizontal line indicates the value of pristine graphene. Inset: The electron–hole asymmetry factor φ of v_F and the bandgap E_g at the Dirac point as a function of θ. Source: Reproduced with permission from Kang et al. [21]. Copyright 2015, American Physical Society.

Table 2.4 The parameters used when calculating the different $V_{pp\pi}$ and $V_{pp\sigma}$.

	$V_{pp\sigma}^{C1-C2}$	$V_{pp\sigma}^{C1-N}$	$V_{pp\pi}^{C1-C1}$	$V_{pp\pi}^{C1-C2}$	$V_{pp\pi}^{C2-N}$	$V_{pp\pi}^{N-N}$
γ (eV)	0.48	0.48	−2.70	−2.70	−2.70	−2.70
a (Å)	3.250	3.250	1.418	1.460	1.335	1.335
q	2.218	5.518	2.218	2.218	5.518	5.518

is relevant, but in the graphene/NHG heterostructure, both ppπ and ppσ interactions are important. According to the Slater–Koster formula [12], t_{ij} is given by:

$$t_{ij} = n^2 V_{pp\sigma}(r_{ij}) + (1 - n^2) V_{pp\pi}(r_{ij})$$

where n is the direction cosine along the z direction of the vector $\vec{r}_j - \vec{r}_i = \vec{r}_{ij}$ and $r_{ij} = |\vec{r}_{ij}|$. The functions $V_{pp\pi}$ and $V_{pp\sigma}$ are assumed to have the form $\gamma \exp(q(1 - r_{ij}/a))$. In the graphene/NHG heterostructure, there are three types of atoms: C1 (C atoms in the graphene layer), C2 (C atoms in the NHG layer), and N. In our calculations, taking the onsite energy of C1 as zero, the onsite energy of C2 and N is −0.27 and −0.70 eV, respectively. For each heterostructure, a constant energy shift is added to all the onsite energies to make the energy of the Dirac point zero. The parameters used for calculating the different $V_{pp\pi}$ and $V_{pp\sigma}$ are listed in Table 2.4. The cutoff radius is chosen as 6 Å, which is sufficient to obtain converged results. Figure 2.13a shows the band structure of the (7,3)–(1,2) G/NHG and graphene calculated by the tight-binding model, and the results agree well with those obtained from DFT.

Here, we consider different Moiré-patterned G/NHG with θ ranging from 0° to 30° because 60°−θ and θ create similar Moiré patterns and they result in the same velocity renormalization. The interlayer spacing is fixed at 3.25 Å as in the (7,3)–(1,2) structure. In all structures studied, the lattice mismatch between graphene and NHG is less than 0.5%. The bandgap opening at the Dirac point is always negligible. As shown in the inset of Figure 2.14c, its order of magnitude is 10^{-4} eV and decreases with decreasing twist angle. The linear band dispersion is well preserved in all structures. The band dispersion around the K point for several selected G/NHG heterostructures is shown in Figure 2.14b. Figure 2.14c presents the calculated Fermi velocity as a function of θ. The following results are found: (i) There is always some electron–hole asymmetry in v_F. When θ is small, the asymmetry factor φ increases rapidly with decreasing θ, as shown in the inset of Figure 2.14c. It reaches a maximum of 13% at $\theta = 0°$. (ii) For $\theta > 15°$, v_F is almost independent of θ, but the value is only about 80% of that of pristine graphene, indicating that velocity renormalization occurs even for large θ. In bilayer graphene, when θ is large, the Moiré pattern periodicity is small, and the pattern-induced potential is almost uniform. As a result, the renormalization effect is weak. However, the NHG has a special crystal structure, with large holes distributed periodically on the monolayer. The potential above the holes is very different from that above other regions. In the G/NHG heterostructures, these holes create nonuniform potentials with a periodicity of a_0^N, which is applied to the graphene layer. For large θ, although the size of the Moiré pattern is small, the velocity renormalization is still significant because of the hole-induced potentials. (iii) For $\theta < 15°$, v_F decreases when the angle decreases. It is known that bilayer graphene exhibits a similar behavior [27, 39]. Moreover, for the graphene bilayer, in the case that θ is extremely small, the size of the Moiré pattern becomes infinite because there is no lattice mismatch between the two graphene layers. This leads to a very strong velocity renormalization, and v_F is reduced to zero [27, 46]. Such a behavior does not appear in the G/NHG heterostructure. When $\theta = 0°$, v_F is reduced by a large extent but not to zero. It is about 30% of that of pristine graphene. Because of the lattice mismatch between graphene and NHG, even at $\theta = 0°$, the Moiré pattern size is still finite, with a periodicity of 86 Å. Therefore, v_F is not reduced to zero as in the graphene bilayer.

References

1 Kang, J., Tongay, S., Zhou, J. et al. (2013). *Appl. Phys. Lett.* 102: 012111.
2 Kang, J., Li, J., Li, S.-S. et al. (2013). *Nano Lett.* 13: 5485–5490.
3 Pan, Y., Folsch, S., Nie, Y. et al. (2018). *Nano Lett.* 18: 1849–1855.
4 Hohenberg, P. and Kohn, W. (1964). *Phys. Rev.* 136: B864–B871.
5 Kohn, W. and Sham, L.J. (1965). *Phys. Rev.* 140: A1133–A1138.
6 Ceperley, D.M. and Alder, B.J. (1980). *Phys. Rev. Lett.* 45: 566–569.
7 Perdew, J.P., Burke, K., and Ernzerhof, M. (1996). *Phys. Rev. Lett.* 77: 3865–3868.
8 Heyd, J., Scuseria, G.E., and Ernzerhof, M. (2003). *J. Chem. Phys.* 118: 8207–8215.
9 Wang, L.-W., Zhao, Z., and Meza, J. (2008). *Phys. Rev. B* 77: 165113.
10 Hedin, L. (1965). *Phys. Rev.* 139: A796–A823.
11 Hybertsen, M.S. and Louie, S.G. (1986). *Phys. Rev. B* 34: 5390–5413.
12 Slater, J.C. and Koster, G.F. (1954). *Phys. Rev.* 94: 1498–1524.
13 Brandbyge, M., Mozos, J.-L., Ordejón, P. et al. (2002). *Phys. Rev. B* 65: 165401.
14 Wang, J.-S., Wang, J., and Lü, J.T. (2008). *Eur. Phys. J. B* 62: 381–404.
15 Yue, Q., Chang, S., Kang, J. et al. (2013). *J. Phys. Chem. C* 117: 14804–14811.
16 Li, G., Li, Y., Liu, H. et al. (2010). *Chem. Commun.* 46: 3256–3258.
17 Baughman, R.H., Eckhardt, H., and Kertesz, M. (1987). *J. Chem. Phys.* 87: 6687–6699.
18 Narita, N., Nagai, S., Suzuki, S., and Nakao, K. (1998). *Phys. Rev. B* 58: 11009–11014.
19 Kang, J., Li, J., Wu, F. et al. (2011). *J. Phys. Chem. C* 115: 20466–20470.
20 Mahmood, J., Lee, E.K., Jung, M. et al. (2015). *Nat. Commun.* 6: 6486.
21 Kang, J., Horzum, S., and Peeters, F.M. (2015). *Phys. Rev. B* 92: 195419.
22 Mak, K.F., Lee, C., Hone, J. et al. (2010). *Phys. Rev. Lett.* 105: 136805.
23 Novoselov, K., Jiang, D., Schedin, F. et al. (2005). *Proc. Natl. Acad. Sci. U.S.A.* 102: 10451.
24 Geim, A.K. and Grigorieva, I.V. (2013). *Nature* 499: 419–425.
25 Yang, W., Chen, G., Shi, Z. et al. (2013). *Nat. Mater.* 12: 792–797.

26 Ponomarenko, L., Gorbachev, R.V., Yu, G.L. et al. (2013). *Nature* 497: 594–597.
27 Trambly de Laissardiere, G., Mayou, D., and Magaud, L. (2010). *Nano Lett.* 10: 804–808.
28 Tran, K., Moody, G., Wu, F. et al. (2019). *Nature* 567: 71–75.
29 Terrones, H., Lopez-Urias, F., and Terrones, M. (2013). *Sci. Rep.* 3: 1549.
30 Kou, L., Frauenheim, T., and Chen, C. (2013). *J. Phys. Chem. Lett.* 4: 1730–1736.
31 Grimme, S. (2006). *J. Comput. Chem.* 27: 1787–1799.
32 Sachs, B., Wehling, T.O., Katsnelson, M.I., and Lichtenstein, A.I. (2011). *Phys. Rev. B* 84: 195414.
33 Lin, Q., Smeller, M., Heideman, C.L. et al. (2010). *Chem. Mater.* 22: 1002–1009.
34 Kosmider, K. and Fernandez-Rossier, J. (2013). *Phys. Rev. B* 87: 075451.
35 Ramasubramaniam, A., Naveh, D., and Towe, E. (2011). *Phys. Rev. B* 84: 205325.
36 Tongay, S., Zhou, J., Ataca, C. et al. (2012). *Nano Lett.* 12: 5576–5580.
37 Wang, L.-W. and Zunger, A. (1994). *J. Chem. Phys.* 100: 2394–2397.
38 MacDonald, A.H. and Bistritzer, R. (2011). *Nature* 474: 453–454.
39 Luican, A., Li, G., Reina, A. et al. (2011). *Phys. Rev. Lett.* 106: 126802.
40 Neek-Amal, M. and Peeters, F.M. (2010). *Phys. Rev. B* 82: 085432.
41 Park, C.-H., Yang, L., Son, Y.-W. et al. (2008). *Nat. Phys.* 4: 213.
42 Fan, Y., Zhao, M., Wang, Z. et al. (2011). *Appl. Phys. Lett.* 98: 083103.
43 Zhong, X., Yap, Y.K., Pandey, R., and Karna, S.P. (2011). *Phys. Rev. B* 83: 193403.
44 Xue, J., Sanchez-Yamagishi, J., Bulmash, D. et al. (2011). *Nat. Mater.* 10: 282.
45 Barbier, M., Vasilopoulos, P., and Peeters, F.M. (2010). *Phys. Rev. B* 81: 075438.
46 Trambly de Laissardiere, G., Mayou, D., and Magaud, L. (2012). *Phys. Rev. B* 86: 125413.

3

Tuning the Electronic Properties of 2D Materials by Size Control, Strain Engineering, and Electric Field Modulation

Precise regulation and control of the electronic properties of materials are significant for a wide range of applications. In the field of traditional materials, complex techniques are required to accurately tune the atomic compositions and structures. In contrast, there are number of ways to tune the physical properties of 2D materials owning to their reduced dimension and special geometrical configurations. In this chapter, we will focus on three types of mostly used methods: size control, strain engineering, and electric field modulation.

3.1 Size Control

As the first mechanically exfoliated 2D material, graphene [1] has been intensively and extensively studied because of its extraordinary and versatile physical properties that are expected to provide great possibilities for next-generation nanoelectronics. However, the nature of zero bandgap severely limits its applications in semiconductor-based electronic devices. In order to achieve a sizable bandgap, various methods have been applied to open the bandgap of graphene, among which the most effective way is to decrease the dimension of graphene into one-dimensional nanoribbons by cutting the exfoliated graphene or by patterning the grown graphene. Graphene can be tailored into nanoribbons with two different edge terminations, namely, an armchair graphene nanoribbon (AGNR) and a zigzag graphene nanoribbon (ZGNR), as illustrated in Figure 3.1. The former one with armchair-shaped edges on both sides is denoted by the number of dimer lines (N_a) for the ribbon width, and the latter one with zigzag-shaped edges on both sides is labeled by the number of zigzag chains (N_z). Upon reducing the dimension, the graphene nanosheet undergoes a semimetal-to-semiconductor transition [2]. The bandgaps of graphene nanoribbons (GNRs) vary in different tendencies with the shapes of the edges. AGNRs show three different categories with the gap size hierarchy of $G_{3n+1} > G_{3n} > G_{3n+2}$ ($\neq 0$) (where n is a positive integer), as illustrated in Figure 3.2. All AGNRs are nonmagnetic direct-bandgap semiconductors with the valence band maximum (VBM) and conduction band minimum (CBM) located at $kd_a = 0$. Moreover, the gap size of AGNRs is characterized by $G_{N_a} \sim w_a^{-1}$ (where w_a is the ribbon width), mainly resulting from quantum confinement effect. Another crucial factor that is responsible for the

Two-Dimensional Semiconductors: Synthesis, Physical Properties and Applications,
First Edition. Jingbo Li, Zhongming Wei, and Jun Kang.
© 2020 Wiley-VCH Verlag GmbH & Co. KGaA. Published 2020 by Wiley-VCH Verlag GmbH & Co. KGaA.

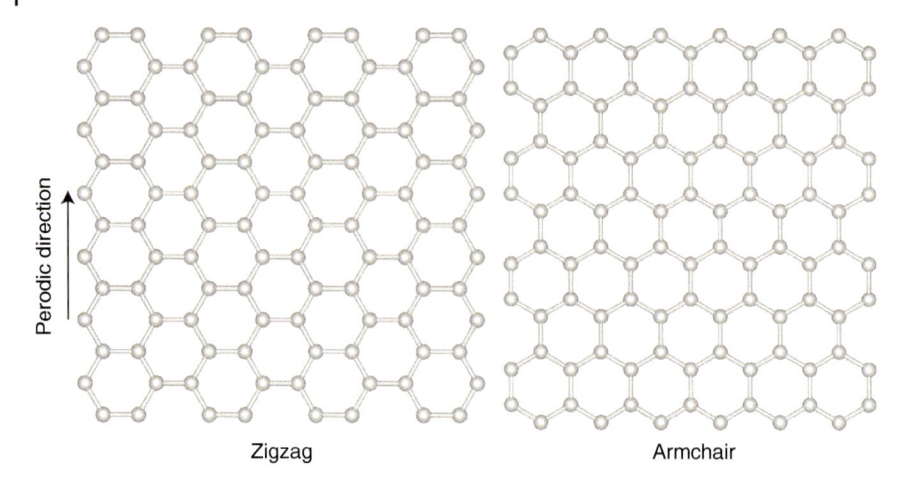

Figure 3.1 Atomic structures of (a) zigzag and (b) armchair graphene nanoribbons.

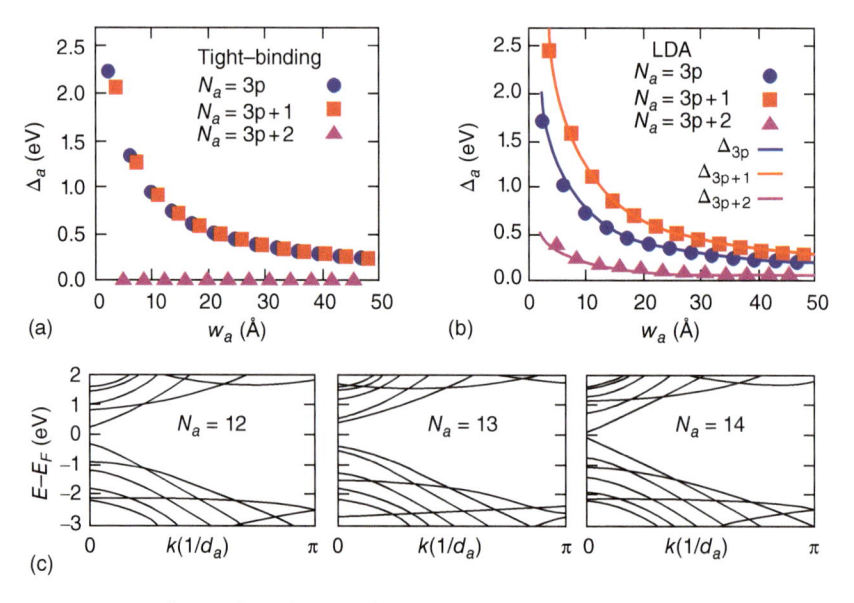

Figure 3.2 Bandgaps of armchair graphene nanoribbons as a function of width calculated by (a) tight-binding method and (b) first-principles calculations. (c) Band structures of armchair graphene nanoribbons with a width of 12, 13, and 14. Source: Reprinted with permission from Son et al. [2]. Copyright 2006, American Physical Society.

semiconducting behavior of AGNRs is the edge effects. The carbon atoms of AGNRs at the edges are saturated by hydrogen atoms; therefore, the σ bonds between the edge carbon atoms are different from those between the middle carbon atoms.

Consequently, the interatomic distances of edge atoms are shortened as compared to those in the middle, thus increasing the hopping integral between the π-orbitals of carbon atoms at the edges. This effect is neglected in tight-binding

(TB) mode, based on which the gaps of the $(3n+2)$-AGNRs are always closed (Figure 3.2a), thus clearly demonstrating the importance of edge effects on the semiconducting behavior in AGNRs. ZGNRs also have direct bandgaps, yet with the VBM and CBM located near $kd_z = \pi$. Particularly, ZGNRs have a magnetic insulating ground state with ferromagnetic coupling at each edge while antiferromagnetic ordering between two edges, which is absent in AGRNs. Without considering the spins, the eigenstates contributed by the edge states of ZGRNs form degenerate flat bands near Fermi level, resulting in a large density of states at Fermi level and infinitesimally small on-site repulsion that could make ZGNRs magnetic at the edges. Similar to the ionic potential difference between B and N atoms that generates the bandgap in the boron nitride (BN) monolayer, the opposite spin states on the opposite edges occupy different sublattices, which introduces exchange potential difference between the two different sublattices on the honeycomb hexagonal lattice and opens the bandgap of ZGNRs. In addition, as the strength of the staggered potential in the middle of the ribbon decays with the width, the bandgaps of ZGNRs induced by the potential difference are inversely proportional to the ribbon width.

Graphyne and its family, a new category of carbon allotrope, have attracted much attention because of graphene-like properties showing possible applications as gas separation membranes, energy storage materials, and anode materials in batteries [3–7]. Recently, graphdiyne [8, 9], with two diacetylenic linkages between repeating patterns of carbon hexagons, has been successfully fabricated via a cross-coupling reaction on a Cu substrate using hexaethynylbenzene. As discussed in Chapter 2, 2D graphyne and graphdiyne are predicted to be semiconductors with a small bandgap of around 0.5 eV predicted by PBE [10, 11]. 1D graphyne and graphdiyne nanoribbons, which can be obtained from the carbomerization of graphene or by cutting the 2D counterparts, are all semiconductors with bandgaps decreasing monotonically with the increasing width, irrespective of the edge shape [10, 12]. Moreover, the 1D graphdiyne nanoribbons are more stable than the 2D counterpart in view of the cohesive energies, and the stability decreases with the increasing ribbon width. The electron mobility of graphdiyne nanoribbons could reach 10^5 cm^2/(V s) at room temperature, which is greater than the hole mobility [10]. The antibonding feature of the highest occupied molecular orbital (HOMO) and the lowest unoccupied molecular orbital (LUMO) is responsible for the gap in mobilities. It is found that the antibonding HOMO has more nodes between carbon hexagons and diacetylenic linkages than the bonding LUMO, which makes the coupling strength of the hole to the acoustic phonon much stronger than that of the electron to the acoustic phonon. As a result, the mobilities of the hole are much smaller than the electron. Another type of graphyne that has no hexagonal rings is named α-graphyne [12], as shown in Figure 3.3. Interestingly, the band structures of both α-graphyne monolayer and nanoribbons have remarkably similar features to the counterparts of graphene. Consequently, α-graphyne not only has linear band dispersion for its monolayer but also shows the same bandgap hierarchy as graphene for its nanoribbons. The armchair α-graphyne nanoribbons are nonmagnetic semiconductors with the bandgap value showing a hierarchy as $G_{3n+1} > G_{3n} > G_{3n+2}$ ($\neq 0$). Zigzag α-graphyne nanoribbons

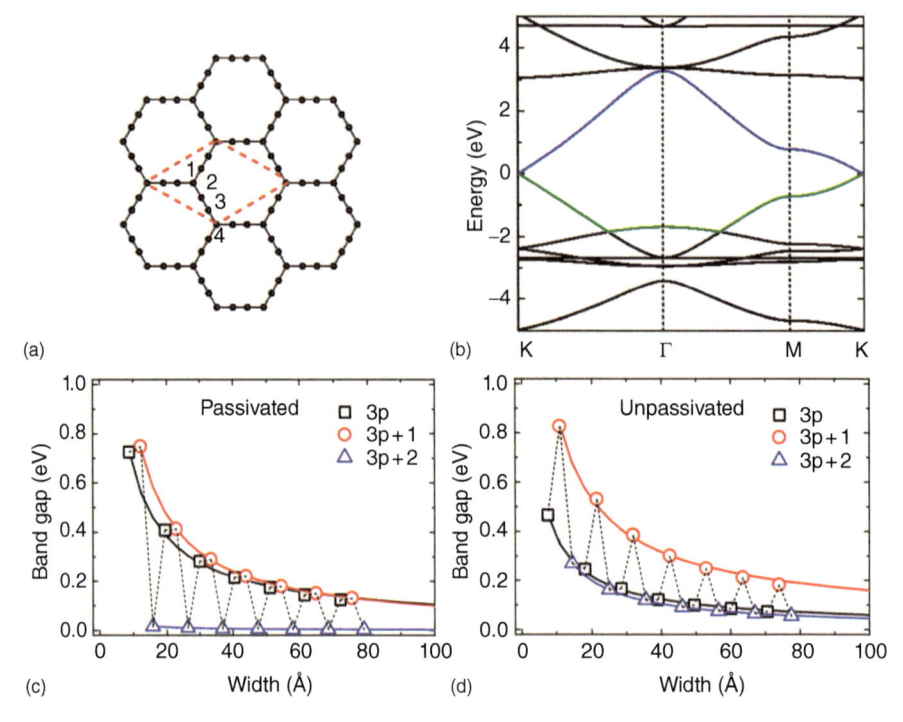

Figure 3.3 (a) Schematic (1) and (4) indicates sp^2 hybridized C atoms, while (2) and (3) indicates sp hybridized C atoms and (b) band structure of 2D α-graphyne sheet. Bandgaps as a function of width for (c) passivated and (d) unpassivated of armchair α-graphyne nanoribbons. Source: Reprinted with permission from Yue et al. [12]. Copyright 2012, AIP Publishing.

possess magnetic insulating ground state with parallel magnetic coupling at one edge and antiparallel magnetic ordering between the two edges.

The effect of quantum confinement is more significant on the physical properties of transition metal dichalcogenides (TMDs). 2D MoS$_2$ has been the most extensively and intensively studied TMD materials because of its inherent bandgap. Besides, an indirect-to-direct bandgap transition occurs when the number of MoS$_2$ layers decreases to a single layer [13], which displays great advantages in applications in future electro-optical devices. A MoS$_2$ monolayer is composed of three atomic layers. A Mo layer sandwiched between two S layers, forming the hexagon with Mo and S atoms alternatively situated at the corners. The monolayers are held together by weak van der Waals (vdW) interaction with the Bernal stacking similar to graphite. It is reported that zigzag MoS$_2$ nanoribbons introduce ferromagnetic interaction on each side and between two sides, mainly arising from the two edge atoms [14], as shown in (Figure 3.4). The magnetism is robust enough to still exist even fully saturating the dangling bonds of edge atoms with H atoms. On the contrary, the armchair MoS$_2$ nanoribbons have nonmagnetic ground states. The zigzag nanoribbons show metallic feature independent of ribbon width and thickness, and the states near Fermi levels are dominated with the 4d electrons of the edge Mo atoms and 3p electrons of edge S atoms. In contrast, the armchair nanoribbons present

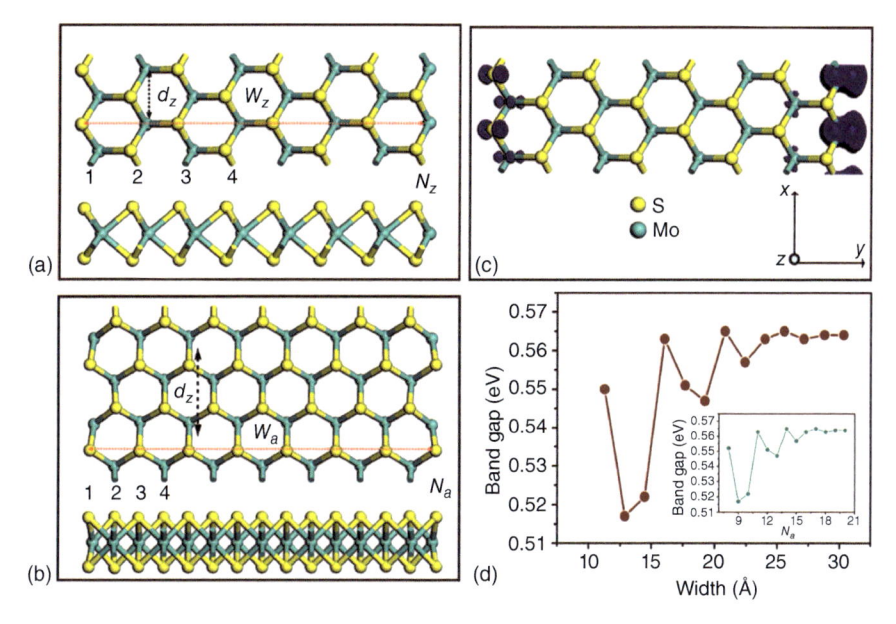

Figure 3.4 Top and side views of atomic structures of (a) zigzag and (b) armchair MoS$_2$ nanoribbons. (c) Spin distribution of zigzag MoS$_2$ nanoribbon. (d) Variation of bandgaps as a function of width for armchair MoS$_2$ nanoribbons. Source: Reprinted with permission from Li et al. [14]. Copyright 2008, American Chemical Society.

semiconducting character with intense oscillation of bandgap for the narrow ribbons. The bandgap finally converges to 0.56 eV for wide ribbons, which is much smaller than that of MoS$_2$ monolayer because of the new flat bands at both valence and conduction band edges, mainly arising from the edge atoms. MX$_2$ (M = Nb and Ta; X = S and Se) has an identical structure to MoS$_2$ but one more valence electron than MoS$_2$; therefore, the MX$_2$ (M = Nb and Ta; X = S and Se) monolayers are all nonmagnetic metals. A similar magnetic coupling was also predicted for MX$_2$ nanoribbons [15]. Most of the armchair-shaped MX$_2$ nanoribbons are nonmagnetic metals, whereas most of the zigzag-shaped MX$_2$ nanoribbons are ferromagnetic metals with magnetic moments located at the edge atoms. Particularly, MX$_2$ nanoribbons undergo metal-to-semiconductor transition when the ribbon width decreases to about 13 and 7 Å for zigzag and armchair edge terminations, respectively. The low-dimensional TMD nanostructures offer a wide range of physical properties spanning from metallic to semiconducting and nonmagnetic to ferromagnetic that are ideal for versatile aspects of applications.

Recently, layered black phosphorus, termed phosphorene [16, 17], has attracted much attention because of its unique electronic properties. Black phosphorus bulk consists of puckered honeycomb layers of phosphorene that are held together by vdW interactions, making the mechanical exfoliation of phosphorene feasible. The bulk form has a direct bandgap of 0.3 eV, and the direct-bandgap feature can be sustained to its 2D form until the monolayer shows a direct bandgap of 1.5 eV. Phosphorene is believed to be superior to graphene

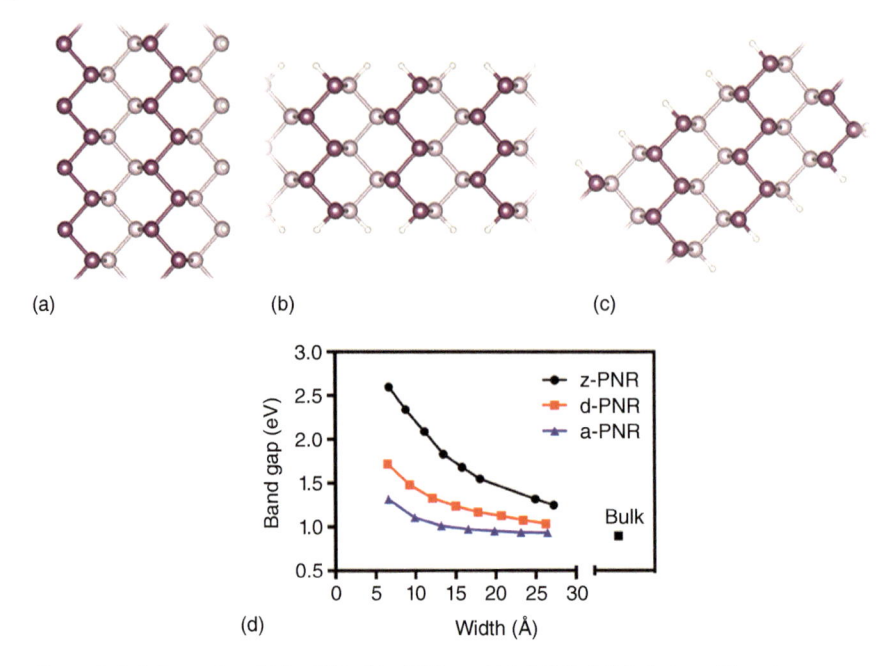

Figure 3.5 Schematics of (a) z-PNR, (b) a-PNR, and (c) d-PNR. (d) Variation of bandgap as a function of ribbon width of phosphorene. Source: Reprinted with permission from Han et al. [18]. Copyright 2014, American Chemical Society.

and TMDs in that it combines the proper bandgap and high mobilities together. As shown in Figure 3.5, three categories of nanoribbons along typical orientations can be tailored in phosphorene: armchair ribbon (a-PNR) along [100], zigzag ribbon (z-PNR) along [010], and diagonal ribbon (d-PNR) along [110] [18]. These nanoribbons have the order of stability of z-PNR > d-PNR > a-PNR. It is reported that zigzag phosphorene nanoribbons show the greatest quantum size effect that tunes the bandgap from 1.4 to 2.6 eV with the width reducing from 26 to 6 Å. Diagonal and armchair phosphorene nanoribbons (PNR) also illustrate a monotonously decreasing bandgap with enhanced ribbon width. The bandgaps of the nanoribbons also follow the order of z-PNR > d-PNR > a-PNR at analogous width.

As shown above, modulations of the ribbon width, edge morphology, and edge functionalization greatly influence the electronic and magnetic properties of nanomaterials that should open up novel and multiple applications of these low-dimensional materials.

3.2 Strain Engineering

Most 2D materials can withstand a large amount of stain as a general feature, endowing them with a large elastic deformation range, which in return makes the modulation of the electronic properties through strain engineering possible. A wide variety of experimental and theoretical works have demonstrated that the electronic and optical gaps of 2D materials are remarkably sensitive to different types of strains [19–23]. Consequently, strain engineering, as a simple yet effective method, has been one of the research hot spots in recent years because of the unique delicate mechanical and electronic coupling in 2D materials. Besides, different types of intrinsic strains exist commonly in 2D materials, such as corrugations, distortion, lattice mismatch with substrates, etc. [24–26]. Various external strains can also be manually induced and precisely controlled in nanomaterials by many advanced techniques. For example, uniform strains can be applied to the layered materials from flexible substrate, on which the layered materials are fabricated [19]. Also, an atomic force microscope (AFM) tip can push the 2D materials on a substrate with precise strains [23], a piezoelectric substrate can bring strains to the materials by applying a bias voltage [27], and the mismatch generated by thermal expansion effect of the substrate can induce strains in layered materials [28, 29]. As the strongest two-dimensional material ever measured, graphene has an in-plane stiffness of 340 ± 50 N/m, corresponding to a Young's modulus of 1 TPa and an intrinsic strength of 130 GPa [23]. Strikingly, it can sustain reversible elastic tensile strain up to 25%, far exceeding that in the silicon of ~1.5% [30]. However, a compressive strain of only less than 0.1% can be applied to a freestanding graphene; otherwise, ripples and wrinkling will be formed to relieve the stress, showing an extreme asymmetry with the tensile strain [31]. However, there is no Dirac shift or bandgap opening under the biaxial strain as the symmetry is well preserved, except for the variations of the slope of the linear dispersion and the Fermi velocity [32]. Strain can also soften the optical phonon branches, which will induce variations in Raman spectra [19, 33, 34]. As is well known, graphene peaks in the Raman spectra: one is the G band at 1580 cm^{-1}, originating from the doubly degenerate E_{2g} phonon mode at the Brillouin zone center, and the other is the 2D band at ~2700 cm^{-1} [35], arising from the four-step double-resonance Raman scattering taking place between the Dirac cone at the K point and its three nearest neighbors at K′ points. As shown in Figure 3.6, red shift and splitting into two single bands are observed when graphene is under uniaxial strain [19, 20, 29]. The uniaxial strain makes the sixfold and threefold rotational symmetries of graphene disappear; therefore, the phonon mode splits into two singlet modes, denoted as G$^+$ and G$^-$ peaks [19]. Also, the 2D band splits into 2D$^+$ and 2D$^-$ under uniaxial strain, which present a red shift with increasing strain. These changes in 2D peaks are stemming from the coupling effects of the Dirac cone shifting and the anisotropic phonon softening under uniaxial strain. Under biaxial strain, there is no splitting of G band and 2D band because of no symmetry loss; only red shifts occur because of the softening of corresponding phonon modes. Remarkably, it is predicted that a biaxial strain of 16.5% combing with a charge doping of ~10^{14} cm^{-2} can bring superconducting character to graphene at a critical temperature of higher than 10 K [30, 34]. This transition is mainly due to the enhanced density of states near the Fermi level from charge doping and strengthened electron–phonon coupling under high strain. For a two-dimensional graphyne sheet, the bandgap is found to be continuously modulated from 0 to 1.3 eV with monotonously increasing feature under the biaxial strain of −10% to 15% [11]. The variation of the bandgap is ascribed to the change in the width of p_z π and p_z π* bands. Under the homogenously tensile

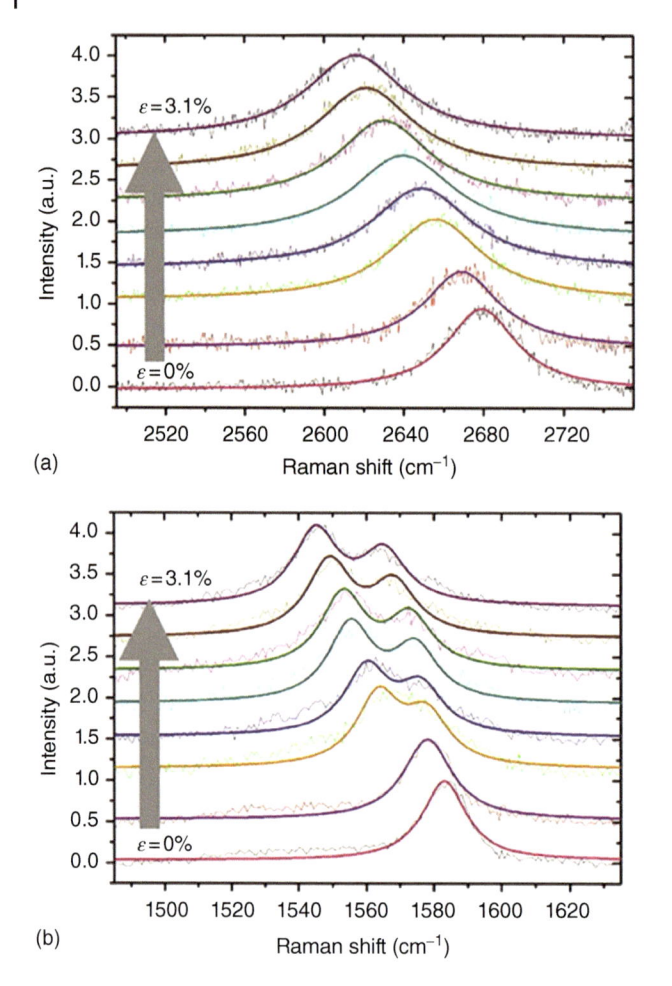

Figure 3.6 Evolutions of the spectra of (a) 2D and (b) G bands of graphene under strain. Source: Reprinted with permission from Huang et al. [19]. Copyright 2009, National Academy of Sciences.

strain, the increasing distance between C atoms leads to the reduction of the overlap of p_z states between C atoms, narrowing the width of p_z π and p_z π^* bands. Consequently, the VBM moves downward, and the CBM moves upward, and the bandgap increases, as illustrated in Figure 3.7. For other members of the graphyne family, the gaps are found to steadily increase with increasing homogeneous tensile strain but decrease with uniaxial tensile or compressive strains. Furthermore, the bandgaps of graphyne and graphyne-3 are always direct and located at either M or S point related with the types of applied tensile strains, while graphdiyne and graphyne-4 always maintain the direct bandgap at Γ point under any types of strains [6].

Semiconducting TMDs are quite sensitive to mechanical strains. For monolayer MoS_2 [36], the bandgap undergoes a descent trend with the increasing strain. Simultaneously, a direct-to-indirect bandgap transition at a strain of 1%

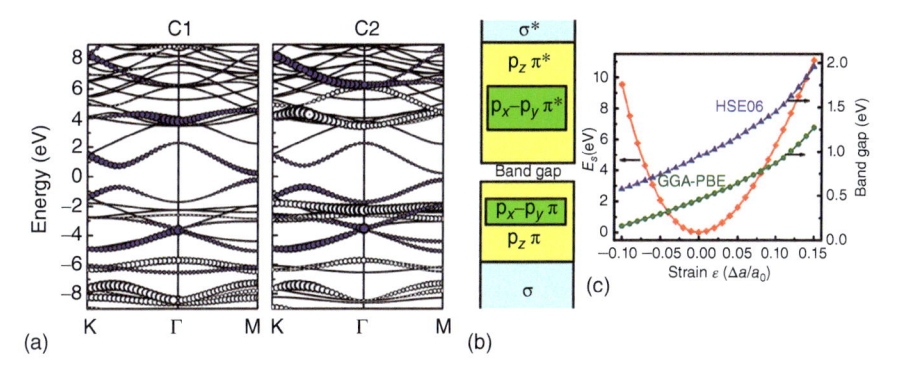

Figure 3.7 (a) Projected band structures of two carbon atoms with sp^2 (C1) and sp (C2) hybridization in graphyne. (b) Schematic of energy levels with respect to different bonding states. (c) Evolution of bandgap and strain energy of graphyne as a function of biaxial strain. Source: Reprinted with permission from Kang et al. [11]. Copyright 2011, American Chemical Society.

and a semiconductor-to-metal transition at a strain of 10% take place. Such involution of the bandgap and structure under strain is due to the changes in orbital overlap integral between the bonding–antibonding states. Specifically, as shown in Figure 3.8, the VBM of the pristine at high-symmetry K point MoS$_2$ is composed of bonding Mo ($d_{xy} + d_{x^2-y^2}$) and S ($p_x + p_y$) orbitals, whereas the state in the same band at Γ point mainly originates from bonding Mo (d_{z^2}) and S (p_z) orbitals. Meanwhile, the CBM of pristine monolayer at K point is mainly contributed by antibonding Mo (d_{z^2}) and S ($p_x + p_y$) orbitals. When the tensile strain is applied, the distance between Mo and S atoms increases, reducing the orbital overlap integral and the splitting between the bonding and antibonding states. Consequently, as the VBM moves up and the CBM moves down, the bandgap decreases. As the Mo ($d_{xy} + d_{x^2-y^2}$) and S ($p_x + p_y$) orbitals are more sensitive to in-plane strains than the Mo (d_{z^2}) and S (p_z) orbitals, the VBM shifts from K point to Γ point under high strains, leading to the direct–indirect bandgap transition. The effective mass of electrons also monotonously decreases with the strain because of the growing flatter band dispersion at CBM induced by the more localized charge distribution under strain. The effective mass of holes shows similar descending tendency with ascending strain with a jump at the critical strain, at which the direct–indirect bandgap transition occurs. In contrast, an indirect-to-direct bandgap transition has been observed in the ZrS$_2$ monolayer when the uniaxial strain is applied along either zigzag or armchair directs because of the similar response difference of the states at high-symmetry reciprocal points to the uniaxial strain.

Besides the in-plane strain, mechanical bending can also control conductivity and Fermi level shift in thin 2D materials. It is found that the in-gap states near the valence bands and the Fermi level pinning stemming from the edge atoms of armchair MoS$_2$ ribbons can be removed by bending [25], as illustrated in (Figure 3.9). A new controllable in-gap state emerges in an armchair phosphorene nanoribbon induced by bending, accompanied with direct–indirect bandgap transition. Biaxial and uniaxial strains in the range of elastic deformation have

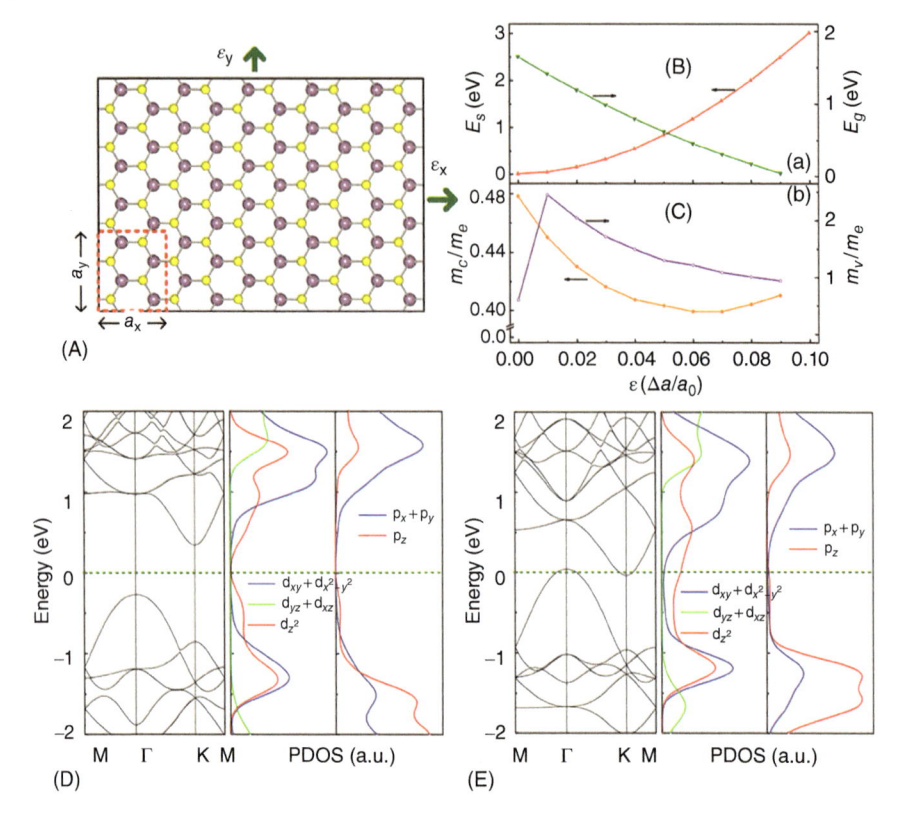

Figure 3.8 (A) Schematic of MoS$_2$ monolayer. Evolutions of (B) bandgap and strain energy and (C) electron and hole effective masses of MoS$_2$ monolayer as a function of strain. Band structures, partial density of states (PDOS) of Mo 4d orbitals and S 3p orbitals for (D) pristine and (E) 5%-strained MoS$_2$ monolayer. Source: Reprinted with permission from Yue et al. [36]. Copyright 2012, Elsevier.

been demonstrated to exert different influences on the electronic structures of phosphorene [37]. The bandgap can be continuously modulated by the strains from 0 to 1.97 eV, which is expected to be promising materials for solar cell and other optical electronic devices.

Strain engineering not only affects the electronic properties of 2D materials but also is able to control their magnetic behavior. Twisting has been found to strongly tune the band structures of GNRs of hundreds of nanometers with arbitrary chirality and induce well-defined levels [38–40], which is similar to the quantization of massive Dirac fermions in a magnetic field of 160 T. This strain-induced pseudomagnetic field makes the electrons localize either at ribbon edges forming the edge current or at the ribbon center forming the snake orbit current, both of them are valley specified. VS$_2$ and VSe$_2$ monolayers are predicted to exhibit magnetic ordering with magnetic moment of 0.434 and 0.584 μ_B, respectively. Moreover, the magnetic moments increase monotonically with increasing isotropic strain from −5% to 5% for these two kinds of monolayers as shown in Figure 3.10 [41]. At the strain of 5%, the magnetic moments

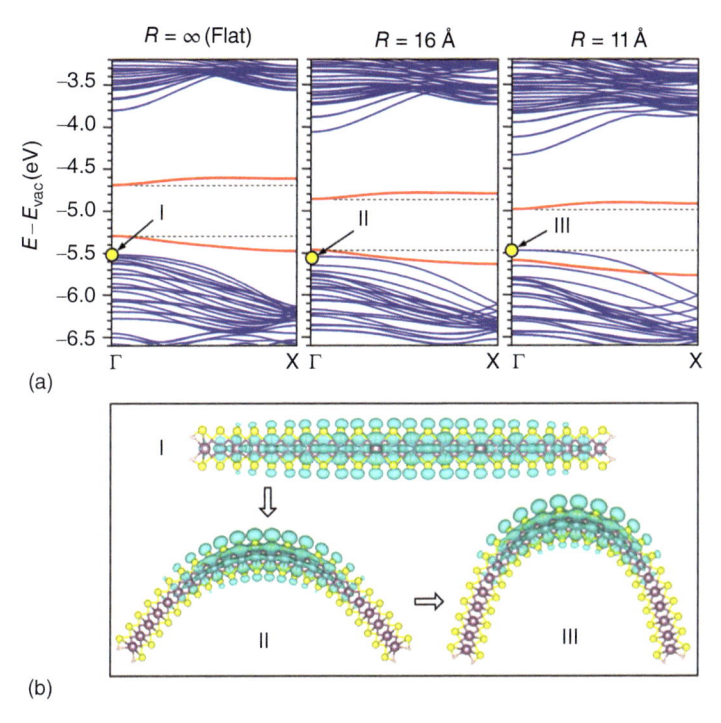

Figure 3.9 (a) Band structures of armchair MoS$_2$ at three different bending curvatures. (b) Isosurfaces of partial charge densities of state I, II, and III corresponding to the three configurations in (a). Source: Reprinted with permission from Yu et al. [25]. Copyright 2016, American Chemical Society.

of VS$_2$ and VSe$_2$ monolayers increase to 1.140 and 1.270 μ_B by enhancement of 187% and 235%, respectively. It is hypothesized that the ferromagnetism in VS$_2$ and VSe$_2$ monolayers arises from two distinct interactions: one is through-bond interaction and the other is through-space interaction. The former one is defined as an atom with an up-spin density introduces a down-spin density on the adjacent atom directly bonded to it. Therefore, the V atoms in these two types of monolayers are parallel spin-polarized, but the V atoms and S or Se atoms are antiparallel spin-polarized according to the through-bond interaction. On the contrary, the latter one is defined as the atom with an up-spin density induces a down-spin density on the nearest-neighboring same atom directly, without mediation of other types of atoms. Therefore, the V atoms are antiparallel spin-polarized between itself, so are S or Se atoms. For pristine VS$_2$ and VSe$_2$ monolayers, the through-bond interaction is stronger than the through-space interaction. Consequently, these two kinds of monolayers show ferromagnetic coupling in their ground states. As both through-bond interaction and through-space interaction are sensitive to the distance between V and S or Se atoms, elongation of the distance induced by the tensile strain gives rise to the reduction in both interactions. Nevertheless, the through-space interaction reduces more quickly than the through-bond interaction, resulting in the relative increase of the ferromagnetic coupling in these two types of monolayers.

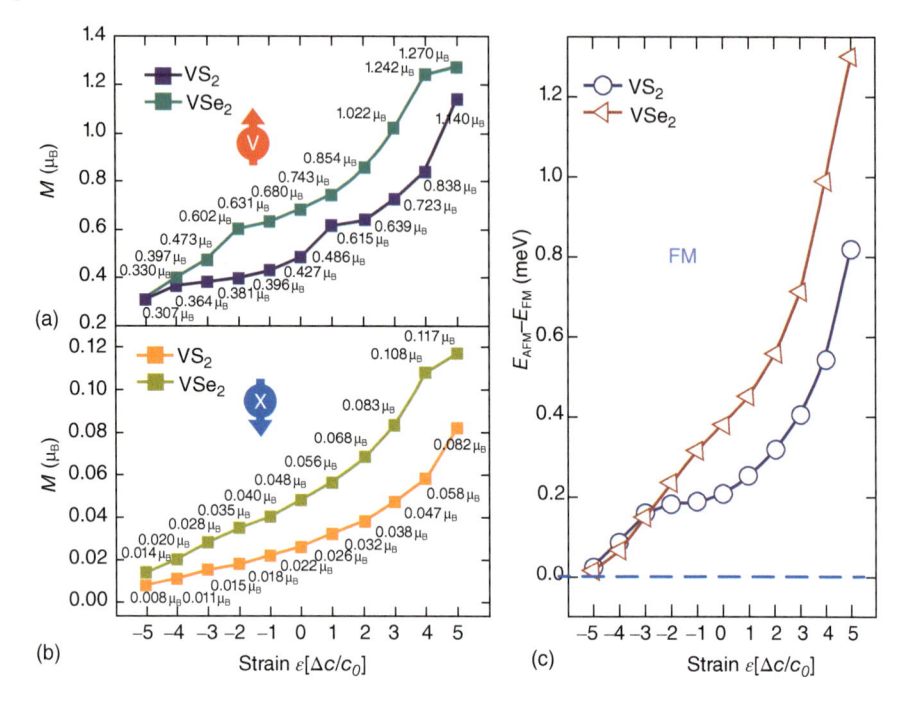

Figure 3.10 Magnetic moment of (a) V atom and (b) X atom as a function of strain. (c) The energy difference between ferromagnetic and antiferromagnetic state of VS_2 and VSe_2 monolayers. Source: Reprinted with permission from Ma et al. [41]. Copyright 2012, American Chemical Society.

Therefore, the ferromagnetism in VS_2 and VSe_2 monolayers is robust enough and even becomes stronger under the external strains. Inspired by the magnetic mechanism of the through-bond and through-space interactions, researchers have examined a series of other TMDs materials to ascertain their magnetic properties. It is reported that the nonmagnetic NbS_2 and $NbSe_2$ monolayers can be magnetized by biaxial tensile strains [42]. Moreover, the induced magnetic moments and the stability of the ferromagnetic behavior can be greatly increased by the increasing tensile strain, even bringing about a half-metallic manner with 100% spin polarization around the Fermi level. Their Curie temperatures are predicted to be 384 and 542 K under 10% stain for NbS_2 and $NbSe_2$ monolayers, respectively, suggesting that these two types of monolayers are suitable for spin applications above room temperature.

As one type of complicated strain, wrinkles have surely been found to affect the electronic and magnetic properties of 2D materials. By introducing local strain through formation of wrinkles on elastomeric substrates, the magnetic, optical, and electronic properties of 2D $ReSe_2$ have been effectively tailored [43]. In contrast to other group of 2D TMDs that are sensitive to the number of layers in every aspect of physical properties, the multilayers (≥ 2) of 2D $ReSe_2$ behave like monolayer owing to much weaker interlayer coupling. Moreover, $ReSe_2$ as well as ReS_2 does not show indirect-to-direct bandgap transition from multilayer to monolayer but possesses strong in-plane anisotropy because of the distorted 1T crystal structural with quasi-1D Re–Re metal chains [44]. Notably, ReSe$_2$ turns magnetic at the strained regions after the formation of wrinkles, confirmed by the magnetic force microscopy (MFM) detection and density functional theory (DFT) calculations [43]. Under the local strain, on the aspect of phonon dispersion, the degenerate in-plane E$_g$-like Raman peak splits into two peaks and the out-of-plane A$_g$-like peak is right-shifted with respect to the unstrained region. The local wrinkle also makes the peak position of the photoluminescence (PL) red shift and the PL intensity increase compared to the unstrained region, implying the changes of the electronic ReSe$_2$ under local strain. DFT calculation affirms that the bandgap is always reduced under tensile strain as experienced by ReSe$_2$ on the top of the wrinkle. On the other hand, the enhancement in integrated PL can be explained by the "funneling effect," which occurs within strained regions because of the nonhomogeneous strain. Consequently, the majority of photogenerated excitons drift to the top of the wrinkle, which shows the maximum strain and smallest bandgap, thereby enhancing the recombination of excitons (Figure 3.11).

To summarize, these findings show that strain engineering can effectively tune the physical properties of nanoscale materials toward creating multifunctional candidates for a wide range of applications in photoelectronic, mechanical, and spintronic devices.

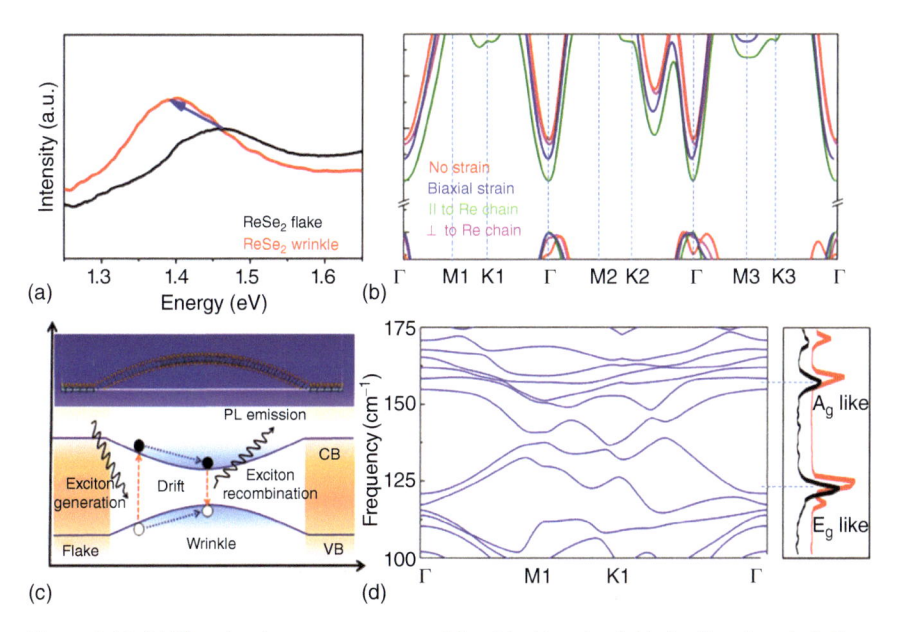

Figure 3.11 (a) Photoluminescence spectra of flat (black) and wrinkle (red) regions in ReSe$_2$ monolayer. (b) Band structures of unstrained and strained ReSe$_2$ monolayers. (c) Schematic of the funnel effect in the wrinkled ReSe$_2$ monolayer. CB, conduction band; VB, valence band. (d) Phonon dispersion of ReSe$_2$ monolayer. (e) Raman spectra of unstrained (black) and strained (red) ReSe$_2$ monolayer. Source: Reprinted with permission from Yang et al. [43]. Copyright 2015, American Chemical Society.

3.3 Electric Field Modulation

As most electronic and optoelectronic devices are designed to be used under the control of external electric field, it is instrumental to understand the effect of electric field on the physical properties of 2D materials. Dielectric constant, ε, is generally used to characterize the response of a material toward the electric field, playing a key role in determining the electron dynamics, such as capacitance, charge screening, and energy storage capability that are related with Coulomb interaction. In 2D crystals, a new series of dielectric features have been observed because of the fact that the Coulomb interactions are confined in a 2D geometry. Based on extensive DFT calculations, it is observed that both the in-plane and out-of-plane dielectric constants are highly tunable by the value of applied electric field [45], as presented in (Figure 3.12). More specifically, both the in-plane and out-of-plane dielectric constants are almost independent (∼1.8 and ∼3, respectively) of the number of layers at a low field strength of less than 0.01 V/Å but increase and become dependent on the number of layers under higher electric field, indicating the stronger modulation of ε with the electric field for thicker 2D materials.

When the vertical electric field is beyond a critical value, graphene becomes unstable and decoupled between layers, therefore offering an instructive method to exfoliate graphene. The driving force for such behavior comes from the linear response of the electrical polarization of the layers under the external field. A similar dielectric response has been observed in 2D MoS$_2$ [46]. At low fields ($E_{ext} < 0.01$ V/Å), the effective dielectric constant of MoS$_2$ keeps a nearly constant value of ∼4, which increases under higher fields depending on the layer numbers. The thicker the layers, the stronger the modulation of dielectric constant under the electric field. Moreover, the tuning ability of dielectric constant with the electric field is stronger on 2D MoS$_2$ than on graphene, suggesting that the MoS$_2$ multilayer is more electrically polarizable than graphene.

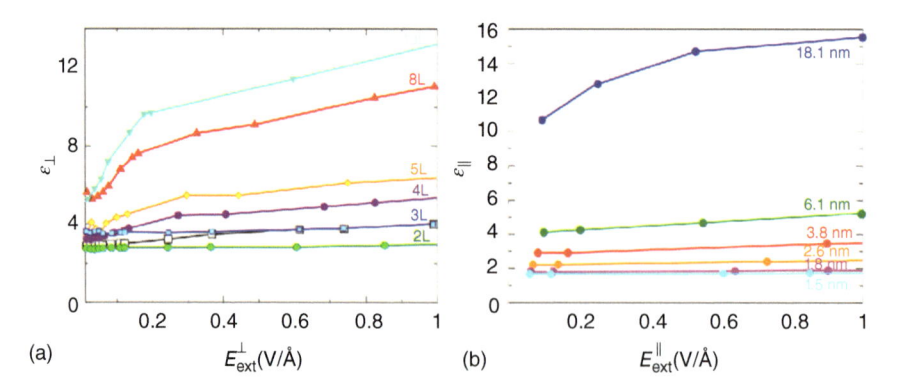

Figure 3.12 (a) Out-of-plane dielectric constant of graphene of different layers as a function of vertical electric field. (b) In-plane dielectric constant of armchair graphene nanoribbons of different widths as a function of parallel electric field. Source: Reprinted with permission from Santos and Kaxiras [45]. Copyright 2013, American Chemical Society.

As is well-known, the plane-wave DFT and pseudopotential methods implemented in most of the codes are based on a supercell geometry, which is composed of periodically repeated slabs and vacuum regions used to eliminate the interactions between the adjacent slabs, in particular for 2D materials. In addition, because of this, it should be noted that the dielectric constant values have been found to strongly depend on supercell size reported for many 2D systems [47–50], even though the convergence of total energy has been fully checked and satisfied against the increase in the size vacuum region. As shown in Figure 3.13, 2D GaS nanosheets have been systematically studied in particular for its dielectric properties [47]. Bulk GaS crystallizes in the β phase with space group P6$_3$/mmc and Bernal stacking, while its monolayer counterpart exhibits honeycomb structure in the form of S–Ga–Ga–S sheet with a space group P$\bar{6}$m2. Uniquely, band inversion occurs in the band structure of GaS with the dimensionality decreases from three to two; as a result, the VBM shifts horizontally from Γ point to two symmetric points nearby, forming the inverted Mexican hat shape. Figure 3.14 depicts the dielectric constant as a function of number of layers. Three sets of vacuum size have been considered: 15, 20, and 30 Å. The dielectric constants demonstrate a linear relationship with the number of layers, while they decrease with the increasing supercell size for the same number of layers. The differences in dielectric constants between diverse vacuum sizes become remarkable with increased thickness. In this respect, the effect of adding GaS layers on the polarization

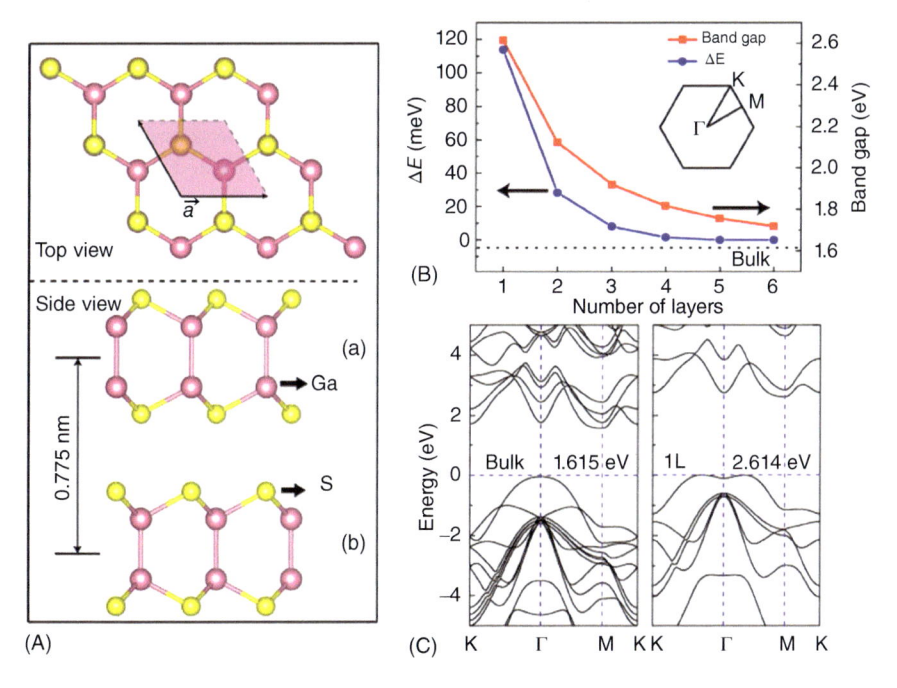

Figure 3.13 (A) Atomic structure of bulk GaS. The rhombic shadow presents the unit cell of GaS. (B) Bandgaps and energy difference between VBM and Γ point as a function of layer thickness. (C) Band structures of bulk and monolayer GaS. Source: Reprinted with permission from Li et al. [47]. Copyright 2015 American Chemical Society.

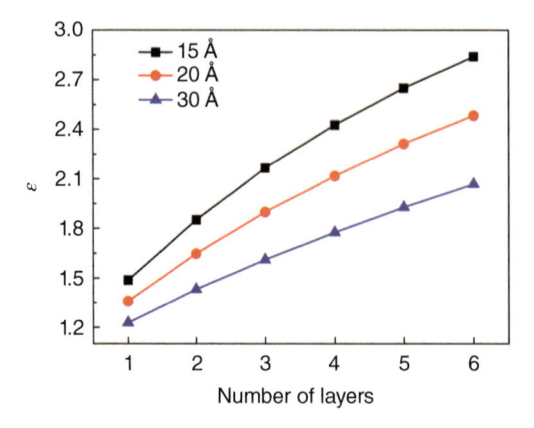

Figure 3.14 Dielectric constant of GaS multilayers as a function of the number of layers. Source: Reprinted with permission from Li et al. [47]. Copyright 2015, American Chemical Society.

and the dielectric response in 2D materials cannot be precisely characterized by the conventional dielectric materials. To solve this nonconvergence between the dielectric constant and vacuum size, polarizability, α, which defines the relationship between the induced dipole moment of an atom and the electric field that generates this dipole moment, has been adopted. For 2D systems, the relationship between the polarizability α and the dielectric constant ε can be expressed as $\alpha = \frac{\Omega}{4}\left(\frac{\varepsilon-1}{\varepsilon}\right)$, where Ω is the supercell volume. Based on this equation, the polarizability of GaS nanosheets with different vacuum spaces can be converged at 15 Å, indicating that this new parameter is independent of the vacuum space when the distance between the periodic GaS stacks is sufficient to prevent electrostatic interactions between periodic images. Also, the slab polarizability increases linearly with the increment of layer numbers, showing the same change tendency with the variation of dielectric constant. Physically, the slab polarizability can be regarded as the effective volume of the slab that acts as a metal that can screen the influence of induced charges under the vertical electric field. The method of calculation slab polarizability has been successfully used to characterize the dielectric behaviors of other 2D materials, such as phosphorene [51], graphene-h-BN heterostructure [52], etc.

Moreover, the external electric field can significantly reduce the bandgaps of most 2D multilayers and 1D nanoribbons and even render them metallic in some instances [12, 37, 47, 48, 53, 54], which is caused by the giant Stark effect (GSE). The relationship between the bandgaps and the applied electric field can be described as $\frac{dE_g}{dE} = -eS$, where e is the electron charge and S is the linear GSE coefficient, which can be used to determine the response of the bandgap under external electric field. The applied electric field not only affects the bandgaps of 2D multilayers but also leads to the charge redistribution because of the induced potential difference. As can be seen in Figure 3.15, the charges at VBM and CBM points are initially distributed over both layers but become localized at the bottom and top layers under the applied electric field of 0.9 V/Å, respectively. In addition, vertical electric field is able to effectively modulate

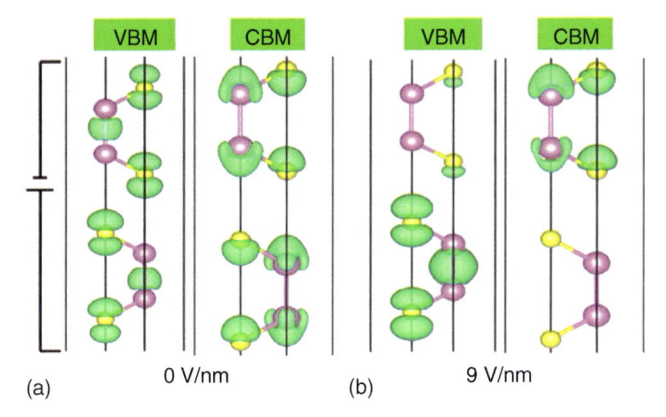

Figure 3.15 Charge density distribution of the VBM and CBM of GaS under external electric field of (a) 0 and (b) 9 V/nm. The isosurfaces are set at 0.005 e/Å^3. Source: Reprinted with permission from Li et al. [47]. Copyright 2015, American Chemical Society.

the charge transfer between the adsorbed molecular and 2D materials, when functioning as gas sensors. DFT calculations show that the charge transfers from monolayer MoS$_2$ to the adsorbed molecular NO or NO$_2$ enhance with the increasing electric field along the direction from the monolayer to the molecule, while reduces with the increment of the opposite electric field [55]. Functionally, the electric field is able to tune the anisotropy of the electronic structure and optical transitions of bilayer black phosphorus with an interlayer twist angle of 90° [56]. It is demonstrated by DFT calculations that a moderate gate voltage can make a hole effective mass 30 times larger along one axis than along the vertical one. Such anisotropy in hole effective mass can be switched under electric field with opposite direct. This field-tunable band structure can generate a switchable optical linear dichroism with the polarization of the lowest-energy optical transitions controllable by external electric field. Furthermore, in-plane electric field can bring in the half-metallicity in graphene ribbons. As discussed above, ZGNRs are semiconductors with ferromagnetic coupled at each edge and antiferromagnetic coupled between two edges, resulting nonmagnetic nature as a whole. With appropriate applied transverse electric field, the edge states nearest to the Fermi level associated with one spin orientation close their bandgap, whereas the edge states associated with the opposite spin orientation widen their gap. As a result, the states with the former spin orientation show metallic behavior, while the states with the latter spin orientation exhibit insulating feature, achieving half-metallic state for ZGNRs [54, 57]. The same mechanism-induced half-metallicity in the zigzag α-graphyne nanoribbons has also been observed under the application of transverse electric field [12].

In summary, all these results give significant and pragmatic information about the modulation and mechanism of the screening and electronic responses in 2D layered materials, which are of great importance in future research and design of novel electronic and optoelectronic devices based on these nanoscale candidates.

References

1 Novoselov, K.S., Geim, A.K., Morozov, S.V. et al. (2004). *Science* 306: 666.
2 Son, Y.W., Cohen, M.L., and Louie, S.G. (2006). *Phys. Rev. Lett.* 97: 216803.
3 Baughman, R.H., Eckhardt, H., and Kertesz, M. (1987). *J. Chem. Phys.* 87: 6687.
4 Narita, N., Nagai, S., Suzuki, S., and Nakao, K. (1998). *Phys. Rev. B* 58: 11009.
5 Haley, M.M. (2008). *Pure Appl. Chem.* 80: 519.
6 Yue, Q., Chang, S.L., Kang, J. et al. (2013). *J. Phys. Chem. C* 117: 14804.
7 Li, Y., Xu, L., Liu, H., and Li, Y. (2014). *Chem. Soc. Rev.* 43: 2572.
8 Li, G.X., Li, Y.L., Liu, H.B. et al. (2010). *Chem. Commun.* 46: 3256.
9 Liu, H., Xu, J., Li, Y., and Li, Y. (2010). *Acc. Chem. Res.* 43: 1496.
10 Long, M., Tang, L., Wang, D. et al. (2011). *ACS Nano* 5: 2593.
11 Kang, J., Li, J.B., Wu, F.M. et al. (2011). *J. Phys. Chem. C* 115: 20466.
12 Yue, Q., Chang, S., Kang, J. et al. (2012). *J Phys. Chem.* 136: 244702.
13 F, M.K., C, L., J, H. et al. (2010). *Phys. Rev. Lett.* 105: 136805.
14 Li, Y., Zhou, Z., Zhang, S., and Chen, Z. (2008). *J. Am. Chem. Soc.* 130: 16739.
15 Li, Y., Tongay, S., Yue, Q. et al. (2013). *J. Appl. Phys.* 114.
16 Liu, H., Neal, A.T., Zhu, Z. et al. (2014). *ACS Nano* 8: 4033.
17 Li, L., Yu, Y., Ye, G.J. et al. (2014). *Nat. Nanotechnol.* 9: 372.
18 Han, X., Stewart, H.M., Shevlin, S.A. et al. (2014). *Nano Lett.* 14: 4607.
19 Huang, M., Yan, H., Chen, C. et al. (2009). *Proc. Natl. Acad. Sci. U.S.A.* 106: 7304.
20 Ni, Z.H., Yu, T., Lu, Y.H. et al. (2008). *ACS Nano* 2: 2301.
21 Kim, K.S., Zhao, Y., Jang, H. et al. (2009). *Nature* 457: 706.
22 Mohiuddin, T.M.G., Lombardo, A., Nair, R.R. et al. (2009). *Phys. Rev. B* 79.
23 Lee, C., Wei, X., Kysar, J.W., and Hone, J. (2008). *Science* 321: 385.
24 Li, Y., Wei, Z., and Li, J. (2015). *Appl. Phys. Lett.* 107: 112103.
25 Yu, L., Ruzsinszky, A., and Perdew, J.P. (2016). *Nano Lett.* 16: 2444.
26 Manzeli, S., Allain, A., Ghadimi, A., and Kis, A. (2015). *Nano Lett.* 15: 5330.
27 Ding, F., Ji, H.X., Chen, Y.H. et al. (2010). *Nano Lett.* 10: 3453.
28 Ferralis, N., Maboudian, R., and Carraro, C. (2008). *Phys. Rev. Lett.* 101: 156801.
29 Yoon, D., Son, Y.W., and Cheong, H. (2011). *Nano Lett.* 11: 3227.
30 Si, C., Sun, Z., and Liu, F. (2016). *Nanoscale* 8: 3207.
31 Zhang, Y. and Liu, F. (2011). *Appl. Phys. Lett.* 99: 241908.
32 Choi, S.-M., Jhi, S.-H., and Son, Y.-W. (2010). *Phys. Rev. B* 81: 081407.
33 Marianetti, C.A. and Yevick, H.G. (2010). *Phys. Rev. Lett.* 105: 245502.
34 Si, C., Duan, W., Liu, Z., and Liu, F. (2012). *Phys. Rev. Lett.* 109: 226802.
35 Ferrari, A.C. (2007). *Solid State Commun.* 143: 47.
36 Yue, Q., Kang, J., Shao, Z. et al. (2012). *Phys. Lett. A* 376: 1166.
37 Li, Y., Yang, S., and Li, J. (2014). *J. Phys. Chem. C* 118: 23970.
38 Guinea, F., Geim, A., Katsnelson, M., and Novoselov, K. (2010). *Phys. Rev. B* 81: 035408.
39 Uchoa, B. and Barlas, Y. (2013). *Phys. Rev. Lett.* 111: 046604.
40 Meng, L., He, W.-Y., Zheng, H. et al. (2013). *Phys. Rev. B* 87: 205405.

41 Ma, Y., Dai, Y., Guo, M. et al. (2012). *ACS Nano* 6: 1695.
42 Zhou, Y., Wang, Z., Yang, P. et al. (2012). *ACS Nano* 6: 9727.
43 Yang, S., Wang, C., Sahin, H. et al. (2015). *Nano Lett.* 15: 1660.
44 Tongay, S., Sahin, H., Ko, C. et al. (2014). *Nat. Commun.* 5: 3252.
45 Santos, E.J. and Kaxiras, E. (2013). *Nano Lett.* 13: 898.
46 Santos, E.J. and Kaxiras, E. (2013). *ACS Nano* 7: 10741.
47 Li, Y., Chen, H., Huang, L., and Li, J. (2015). *J. Phys. Chem. Lett.* 6: 1059.
48 Kozinsky, B. and Marzari, N. (2006). *Phys. Rev. Lett.* 96: 166801.
49 Tobik, J. and Dal Corso, A. (2004). *J Phys. Chem.* 120: 9934.
50 Yu, E.K., Stewart, D.A., and Tiwari, S. (2008). *Phys. Rev. B* 77: 195406.
51 Kumar, P., Bhadoria, B.S., Kumar, S. et al. (2016). *Phys. Rev. B* 93: 195428.
52 Kumar, P., Chauhan, Y.S., Agarwal, A., and Bhowmick, S. (2016). *J. Phys. Chem. C* 120: 17620.
53 Kang, J., Wu, F., and Li, J. (2012). *J. Phys.: Condens. Matter* 24: 165301.
54 Young-Woo Son, M.L.C. and Louie, S.G. (2006). *Nature* 444: 347.
55 Yue, Q., Shao, Z., Chang, S., and Li, J. (2013). *Nanoscale Res. Lett.* 8: 425.
56 Cao, T., Li, Z., Qiu, D.Y., and Louie, S.G. (2016). *Nano Lett.* 16: 5542.
57 Kan, E.J., Li, Z.Y., Yang, J.L., and Hou, J.G. (2007). *Appl. Phys. Lett.* 91: 243116.

4

Transport Properties of Two-Dimensional Materials: Theoretical Studies

4.1 Symmetry-Dependent Spin Transport Properties of Graphene-like Nanoribbons

4.1.1 Graphene Nanoribbons

Graphene-based nanostructures have drawn enormous attention because of their remarkable physical properties and application potential in nanotechnology [1–7]. In particular, graphene nanoribbons (GNRs) exhibit edge magnetism and unique transport properties, which are of interest for spintronics [8–11]. The ground state of zigzag graphene nanoribbon (ZGNR) is found to be antiferromagnetic (AFM) with parallel spin at each zigzag edge and antiparallel spin between two edges [8, 12]. It has been demonstrated that the spin transport behaviors depend on the symmetry of ZGNRs [13–16]. Usually, ZGNRs can be classified into two groups with respect to whether there is a mirror plane [10]. In addition, the two groups of ZGNRs exhibit distinctly different transport behaviors. It is also proposed that ZGNR can have a dual-spin filter effect under finite bias with the up- and down-spin electrons filtered at the counter direction bias [17, 18].

The symmetry-dependent spin transport behaviors can be explored by calculating the current I_σ in systems as a function of the applied external bias V_b [19]:

$$I_\sigma(V_\sigma) = (e/h) \int T_\sigma(E, V_b)[f_l(E, V_b) - f_r(E, V_b)]\, dE$$

where $\sigma = \uparrow$ (spin-up) and $\downarrow$ (spin-down), $T_\sigma(E, V_b)$ is the bias-dependent transmission coefficient, and $f_{l/r}(E, V_b)$ is the Fermi–Dirac distribution function of the left (right) electrode. It is found by Kim et al. that asymmetric ZGNRs behave as conventional conductors, whereas symmetric ZGNRs display a very high resistance [20]. These distinct transport properties of ZGNRs are attributed to the different coupling between the subbands [10]. However, the band structures of asymmetric and symmetric ZGNRs in Figure 4.1 exhibit

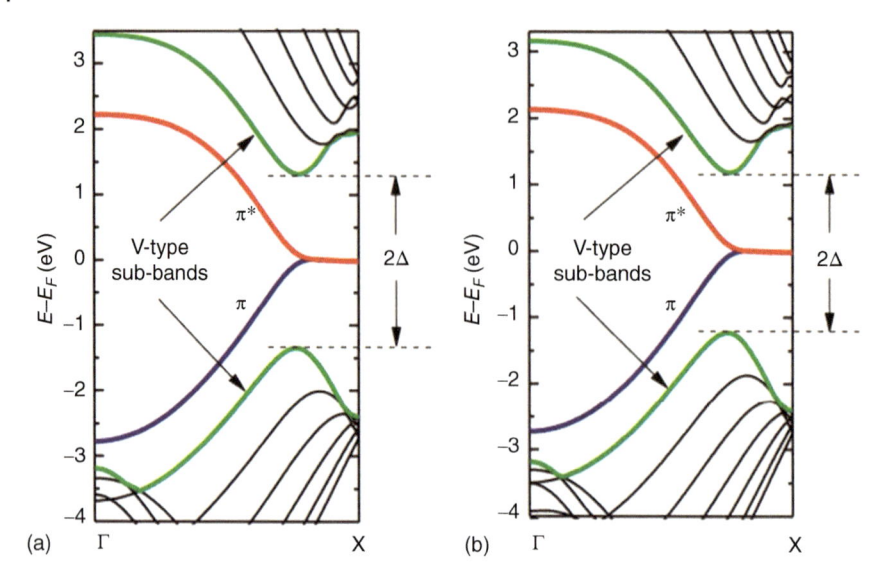

Figure 4.1 (a, b) Band structures around the Fermi level for 7-ZGNR and 8-ZGNR. Source: Reprinted with permission from Li et al. [10]. Copyright 2008, American Physical Society.

very similar features. Many previous works have revealed two flat bands near the Fermi level in their band structures, which are introduced by the edge states. These two edge states mix and form π bonding states and π* antibonding states. Different characteristics are observed in the wave functions of the edge states of symmetric and asymmetric ZGNRs. In accordance with their symmetric geometry, the π (π*) subbands of symmetric ZGNRs have odd (even) parities under σ mirror operation, whereas the edge states of asymmetric ZGNRs have no definite parity because of the absence of the σ mirror operation.

Furthermore, the current through the symmetric ZGNRs can be remarkably enhanced by asymmetric edge terminations. This feature implies that the conductivity of ZGNRs can be controlled by changing their symmetries. Various approaches have been explored to break the spin degeneracy of ZGNRs, such as edge modification [21], application of external electric field [12, 22], and doping [23]. When under a bias voltage, an energy shift in the Fermi level will appear, resulting in an energy window, as shown in Figure 4.2 [10]. This energy window can be divided into two regions: V_{g+} and V_{g-}. As a result, only a π subband exists in the V_{g+} region and a π* subband exists in the V_{g-} region. For symmetric ZGNRs, there is no coupling between the π and π* subbands because of their opposite σ parity. As a result, there is no transmission in the energy window, and then, a conductance gap appears. For asymmetric ZGNRs, the transition of an electron from a π state to a π* state is allowed, which contributes to the transmission.

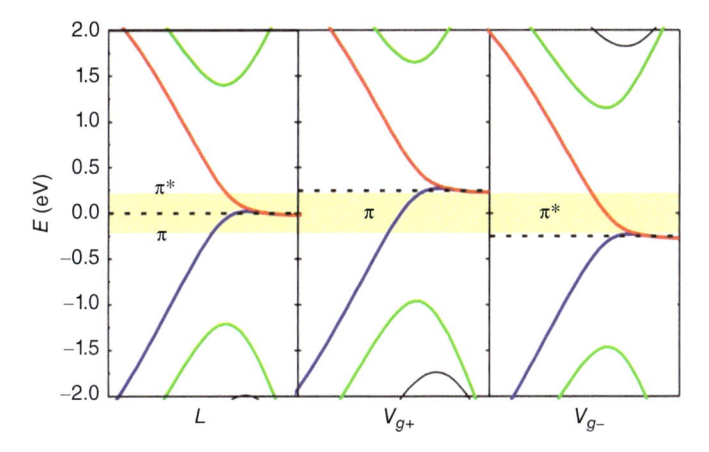

Figure 4.2 Schematic band structures of the left lead and two-gate regions. Source: Reprinted with permission from Li et al. [10]. Copyright 2008, American Physical Society.

As expected, the currents can be enhanced by breaking the symmetry of the system. This can be achieved by passivating both edges of the nanoribbon with different atoms or molecules. For example, if the two edges of ZGNRs are passivated by H and OH, respectively, their electronic structures do not change visibly, but the current under bias voltages increases remarkably. This is attributed to the broken mirror symmetry. As a result, π and π^* states can couple with each other and contribute to the conductance.

4.1.2 Graphyne Nanoribbon

Graphyne is another allotrope of carbon. There is sp and sp^2 hybridization between carbon atoms in graphyne. Because of the existence of acetylenic bonds in its structure, graphyne is expected to have rich electronic and optical properties that are different from those of graphene [24–30]. Specifically, graphyne is found to have a narrow band gap [30]. However, some graphyne derivatives, such as α-graphyne, are gapless semiconductors like graphene [31, 32].

The electronic and transport properties of zigzag α-graphyne nanoribbons (ZαGNRs) have been studied systematically. It is found that the semiconducting ZαGNRs prefer AFM states rather than ferromagnetic (FM) states, which is similar to ZGNRs [33]. Moreover, the energy difference between these two spin states decreases with the nanoribbon width.

To study the symmetry-dependent transport properties of ZαGNRs, symmetric and asymmetric ZαGNRs (Figure 4.3) have been employed in some works [10, 12, 34, 35]. As shown in Figure 4.3 [35], ZαGNRs show a symmetry-dependent transport properties, which is similar to that of ZGNRs. For asymmetric ZαGNRs without σ mirror operation, a linear variation of the current with increasing bias voltage is observed. However, in the case of symmetric ZαGNRs, an energy window appears with zero current under a finite bias is observed.

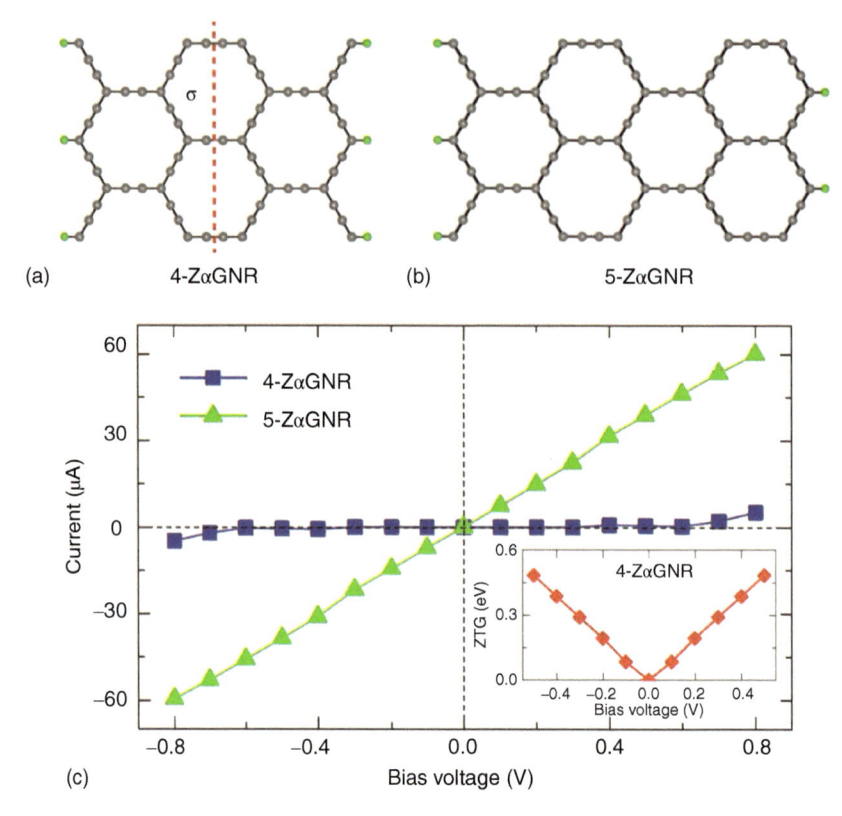

Figure 4.3 Geometric structures of (a) symmetric 4-ZαGNR and (b) asymmetric 5-ZαGNR. (c) The variation of current as a function of bias voltage. The zero transmission gap versus bias for the 4-ZαGNR is given in the inset. Source: Reprinted with permission from Yue et al. [35]. Copyright 2012, American Physical Society.

The physical mechanism underlying the distinct transport properties of ZαGNRs can be understood by exploring their electronic structures and wave functions of the edge states, as show in Figure 4.4. It is found that symmetric and asymmetric ZαGNRs show similar electronic structures with π bonding states and π* antibonding states around the Fermi level. From the wave functions of the π and π* subbands, it can be seen that the π and π* subbands show different characteristics in their wave functions.

Specifically, the π (π*) subband of symmetric ZαGNRs has even (odd) parity under the σ mirror operation. However, for the asymmetric ZαGNRs, there is no definite parity because of the absence of the mirror plane.

When symmetric ZαGNRs are subjected to a bias voltage, the coupling between the π subbands of the left electrode and the π* subbands of the right electrode is forbidden because of their opposite parities. Consequently, there is no electron transmission in ZαGNRs. For asymmetric ZαGNRs, the electron transmission from π subband to π* subband is allowed, leading to the linear current–voltage variation in Figure 4.4.

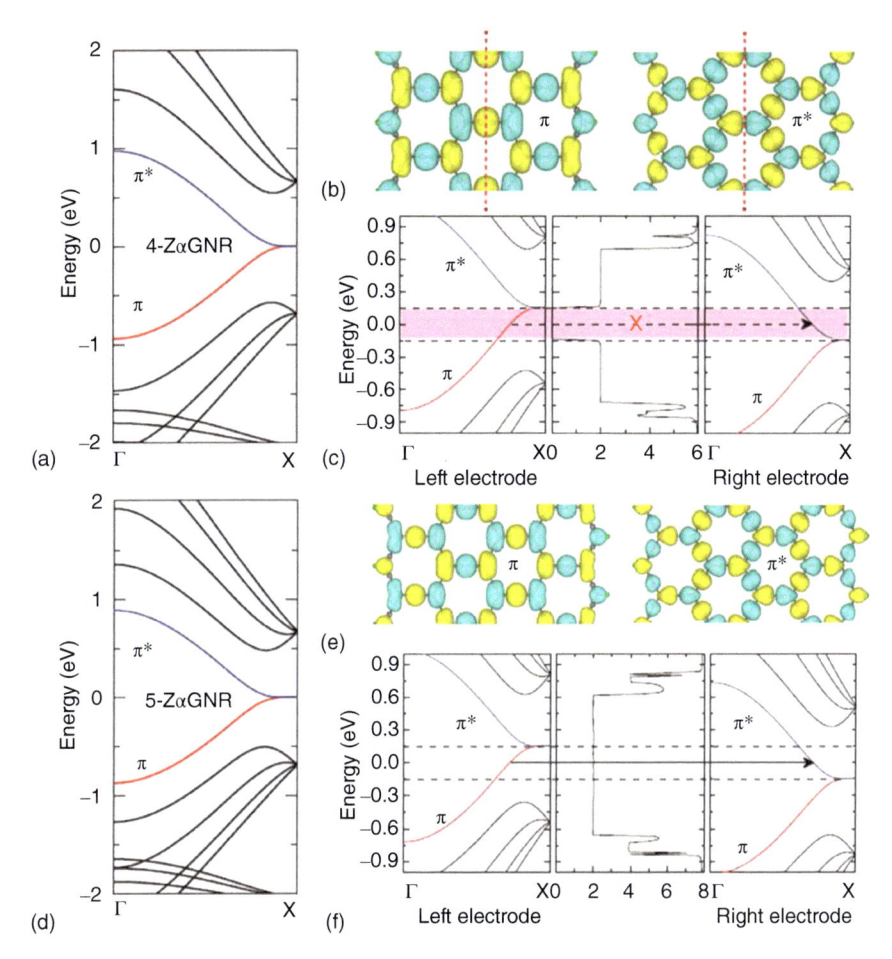

Figure 4.4 (a) Band structure of 4-ZαGNR. (b) Γ-point wave functions of π and π* subbands for 4-ZαGNR. The yellow and blue colors indicate positive and negative signs, respectively. Red dashed lines denote the mirror plane. (c) Band structures for the left and right electrodes, and transmission spectrum for 4-ZαGNR under 0.3 bias voltage. (d–f) The same caption as (a–c) but for 5-ZαGNR. The horizontal dashed lines in (c) and (f) denote the chemical potentials of left and right electrodes. Source: Reprinted with permission from Yue et al. [35]. Copyright 2012, American Physical Society.

4.1.3 Silicene Nanoribbons

Silicene has similar electronic structures to that of graphene with a Dirac cone at the K point. High carrier mobility is demonstrated in silicene. Furthermore, silicene also shows great potential for spintronics application. Zigzag silicene nanoribbons (ZSiNRs) are predicted to have an AFM coupling between its edge magnetic states [36–38]. Kang et al. have calculated the properties of ZSiNRs by nonequilibrium Green's function method and Landauer–Buttiker formula. Symmetry-dependent transport properties are observed in ZSiNRs with the symmetric and asymmetric ZSiNRs showing distinct current–voltage relationships,

Figure 4.5 (a) Band structure of FM 6-ZSiNR. The spin-up and spin-down components are presented in red and blue, respectively. (b) and (c) Spin density for 6-ZSiNR with P and AP configurations under zero bias. The two electrodes have parallel spin configuration in the P case but antiparallel spin configuration in the AP case. Pink and blue surfaces denote the spin-up and spin-down components, respectively. The isosurface corresponds to 0.01 $e/\text{Å}^3$. (d) The I–V curves for the P and AP configurations. The inset is semilogarithmic-scale plot. (e) The spin-up, spin-down, and total magnetoresistance on semilogarithmic scale. (f) and (g) The band structures for left and right electrodes, and the transmission spectrum for P (f) and AP (g) configurations under zero bias. Solid and open circles denote π and π^* states. Solid arrows indicate allowed transmissions, and dashed arrows indicate forbidden transmissions [39]. Source: Reprinted with permission from Yue et al. [35]. Copyright 2012, American Institute of Physics.

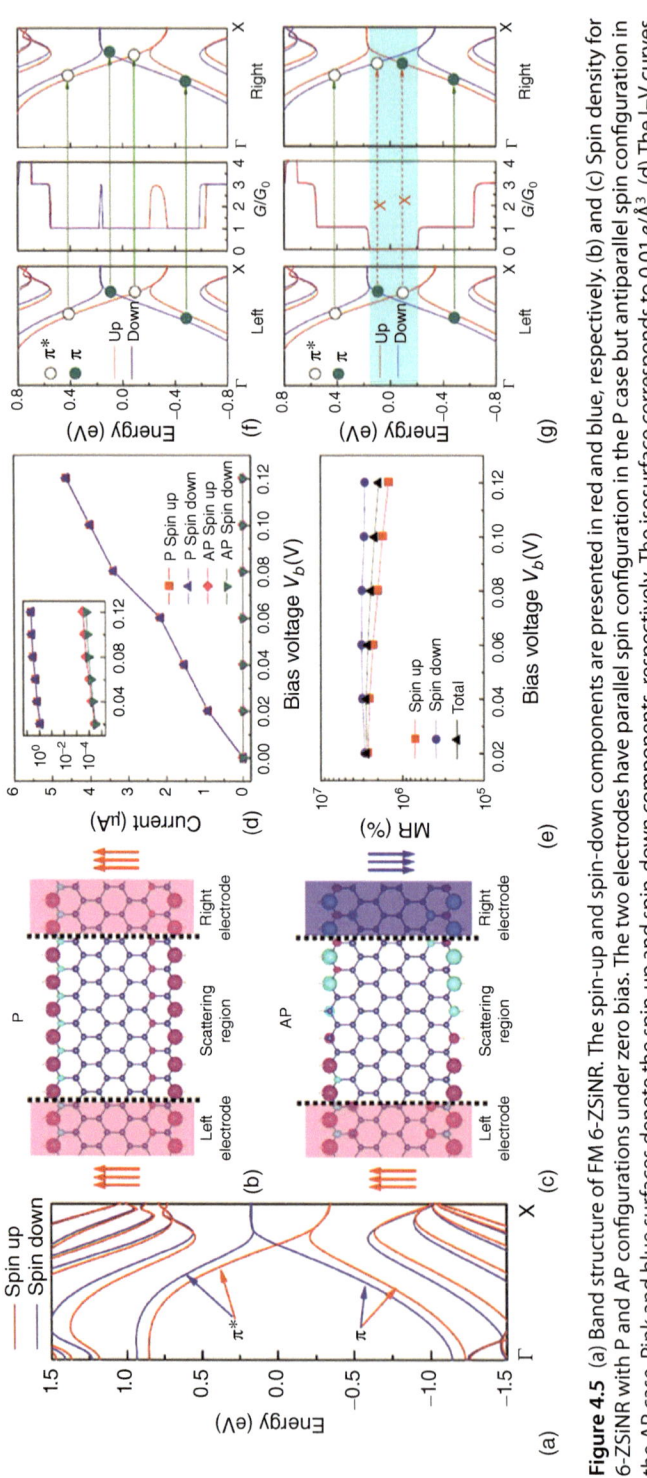

which are similar to those in ZGNRs [39]. These different transport behaviors of symmetric and asymmetric ZSiNRs underlie the different parities of the π and π^* edge states. Different from symmetric ZGNRs, where a σ mirror plane exists, in ZSiNRs, the nearest Si atoms belong to different sublattices, which are not in the same plane because of the buckled structure of silicene. Therefore, ZSiNRs have a lower symmetry than ZGNRs. It has been demonstrated that the c_2 symmetry operation with respect to the centre axis plays an important role in the symmetry-dependent transport properties of ZSiNRs. For symmetric ZSiNRs, under the c_2 symmetry operation, the wave function of the π bonding edge state satisfies $c_2 \varphi_k^\pi(x, y) = \varphi_k^\pi(-x, -y) = \varphi_k^\pi(x, y)$, which indicates that it has even parity. However, the wave function of the π^* antibonding edge state satisfies $c_2 \varphi_k^{\pi*}(x, y) = \varphi_k^{\pi*}(-x, -y) = -\varphi_k^{\pi*}(x, y)$, suggesting a odd parity in this wave function. Because of the opposite parity, the electron transmission from the π band of the left electrode to the π^* band of the right electrode is forbidden, resulting in a conductance gap near the Fermi level. On the other hand, in asymmetric ZSiNRs that have no c_2 cymmetry, the electron transmission from the π band to the π^* band is not limited by parity. As a result, a linear current–voltage variation is observed.

Similar to FM symmetric ZGNRs with a σ mirror plane, magnetoresistance (MR) also can be realized in FM symmetric ZSiNRs with c_2 symmetry. When the left and right electrodes are spin-up polarized (parallel configuration), the current increases linearly with applied bias voltage, as shown in Figure 4.5 [39]. This low-resistance behavior is attributed to the coupling between the spin-up π states of the left electrode and the spin-down π^* states. The transmission is allowed in the case of AP configuration. Although the left and right electrodes are spin antiparalleled (AP configuration), ZSiNRs with c_2 symmetry exhibit MR. In this configuration, only the spin-up π^* states of the left electrode can couple with spin-up π states of the right electrode. As a result, the transmission between them is forbidden because of their opposite parity under the c_2 symmetry. The magnetoresistance effect has great potential application in logic device.

4.2 Charge Transport Properties of Two-Dimensional Materials

4.2.1 Phonon Scattering Mechanism in Transport Properties of Graphene

The transport properties of semiconductors are always limited by different scattering mechanisms including the scattering of phonons, impurities, and electrons. Among all the scattering mechanisms, electron–phonon scattering is of the greatest importance because it determines the ultimate limit of the electronic device performance. The electron–phonon coupling is usually explored by using the deformation potential approximation. Many researchers have applied this theory to study the transport properties of graphene.

Kim et al. have explored the electron–phonon coupling strength in graphene by using first-principles calculations [40]. The electron–phonon interaction matrix elements are calculated by using the density functional perturbation theory (DFPT) with the density functional theory (DFT), in which each phonon is treated as a perturbation of a self-consistent potential created by all electrons and ions [41]. The elements of the electron–phonon interaction matrix is calculated as

$$g_{q,k}^{(i,j)v} = \sqrt{\frac{\hbar}{2M\omega_{v,q}}} \langle j|, k+q|\Delta V_{q,\text{SCF}}^v|i, k \rangle$$

where $|i, k\rangle$ denotes the Bloch eigenstate with wave vector k band index i. E_k^i is the energy eigenvalue. $\Delta V_{q,\text{SCF}}^v$ is the derivative of the self-consistent Kohn–Sham potential with respect to the atomic displacement. In addition, this atomic displacement is characterized as a phonon of branch v with wave vector q and frequency $\omega_{v,q}$. M is a quantity of the atomic mass, which depends on the crystal structure and phonon polarization. Then, according to the matrix elements in the first Brillouin zone, the scattering range can be calculated as Fermi's golden rule:

$$\frac{1}{\tau_k^i} = \frac{2\pi}{\hbar} \sum_{q,v,\pm} |g_{q,k}^{i,v}|^2 \left(N_{v,q} + \frac{1}{2} \pm \frac{1}{2}\right) \delta(E_{k\mp q}^i \pm \hbar\omega_{v,q} - E_k^i)$$

where $\pm$ represents phonon emission and absorption, respectively. $N_{v,q}$ is the phonon occupation number in the Bose–Einstein statistics.

Figure 4.6 shows the calculated scattering rates of graphene monolayer. It is found that the scattering rates of the out-of-plane phonons (acoustic ZA and optical ZO modes) are much smaller than those of the in-plane modes. A linear relationship is observed in the coupling between electrons and in-plane phonons.

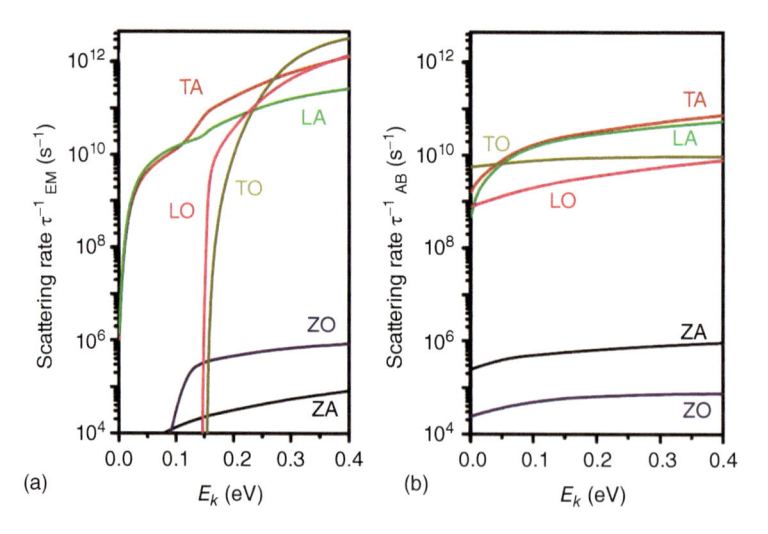

Figure 4.6 Phonon (a) emission and (b) absorption scattering rates at $T = 300$ K as functions of electron energy E_k, as k changes along the K–Γ direction in monolayer graphene. The contributions of all six branches are shown separately. The out-of-plane phonons (ZA and ZO) do not play an important role. At very low electron energies, the scattering with absorption of TO phonons is dominant. Source: Reprinted with permission from Borysenko et al. [40]. Copyright 2010, American Physical Society.

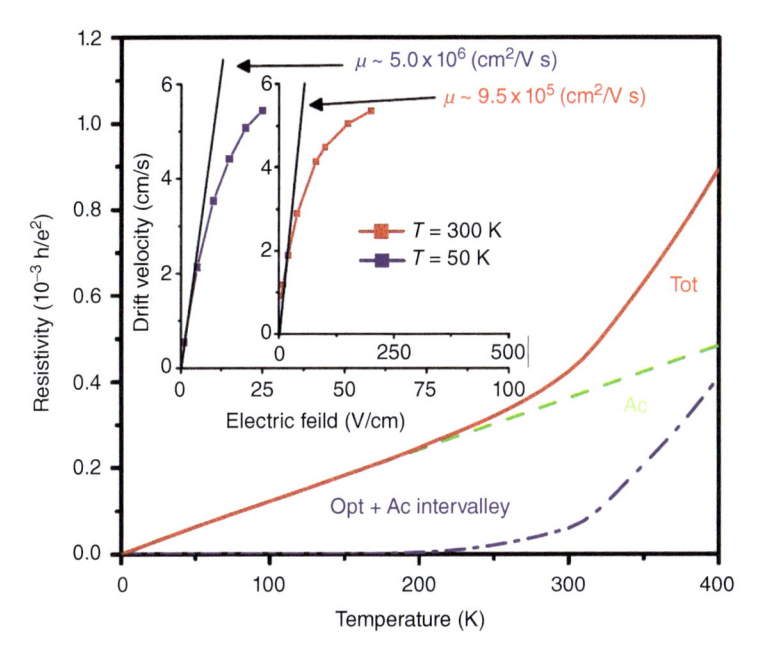

Figure 4.7 Intrinsic resistivity as a function of temperature. At $T < 200$ K, the quasielastic intravalley scattering with in-plane acoustic phonons is dominant (dashed line). At higher temperatures, the contributions of both optical phonons and intervalley scattering by TA and LA modes (dashed-dotted line) lead to an exponential growth of the resistivity. Source: Reprinted with permission from Borysenko et al. [40]. Copyright 2010, American Physical Society.

All the four in-plane modes contribute comparable scattering rates despite their different phonon dispersions. From the calculated scattering rates, the intrinsic transport properties of graphene can be predicted. As shown in Figure 4.7, the coupling of electrons with transverse acoustic (TA) and longitudinal acoustic (LA) phonons exhibit an approximately quasielastic relationship under low temperatures, resulting in a linear dependence of resistivity on temperature. When the temperature is up to 200 K, the optical phonons and intervalley scattering rates increase. From the slope of the velocity field curves in the linear range in Figure 4.7, the low-field mobility can be estimated. A very high mobility of approximately 5×10^6 cm^2/V s is predicted at 50 K. This calculated mobility of graphene is much higher than experimental results because of the absence of other scattering mechanisms in this model.

4.2.2 Phonon Scattering Mechanism in Transport Properties of Transition Metal Dichalcogenides

Transition metal dichalcogenides (TMDs) have attracted enormous attention because of their fascinating electrical, optical, and spintronic properties [42–47]. The transition metals consist of Mo, W, Zr, Hf, or Pt. Chalcogenide atoms are S, Se, or Te. These two-dimensional (2D) materials possess finite band gaps, making them semiconductors with outstanding transport properties. A

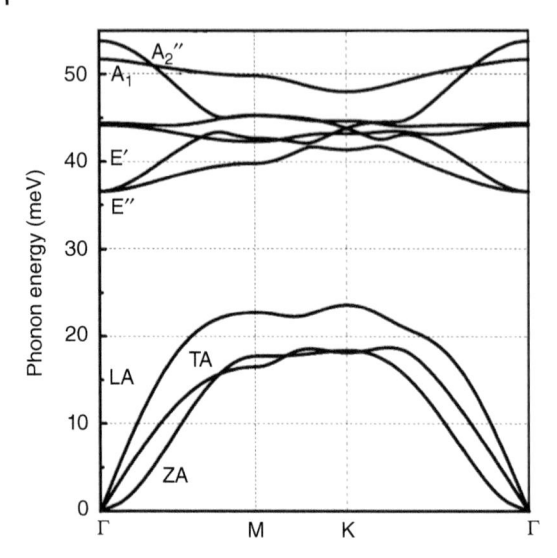

Figure 4.8 Phononic dispersion of monolayer WS_2 in the first Brillouin zone. Source: Reprinted with permission from Jin et al. [55]. Copyright 2014, American Physical Society.

systematical investigation of the transport properties of TMDs is crucial for understanding the technological significance of these materials. Although most of the experimental studies have been carried out under a wide range of conditions [48–51], it is difficult to determine the extrinsic factors that have influence on the current–voltage (I–V) measurements [52, 53]. Theoretical mechanism underlying the transport properties of TMDs is of much importance.

Before theoretically exploring the transport properties of TMDs, the DFT is usually adopted to determine their electronic structures and the phonon spectra [54]. Then, the matrix elements of the electron–phonon interaction can be calculated according to the electronic and phonon structures. The intrinsic transport properties can be evaluated by solving the Boltzmann transport equation. The scattering matrix elements can be calculated by the DFPT along with the electronic and phononic dispersion relations. Some packages within DFT, such as the QUANTUM ESPRESSO and PWmat package, can be used to do the calculations.

Based on the DFT formalism, Kim and co-workers have investigated the intrinsic transport properties of monolayer TMDs [55]. It is found that MX_2 (M = Mo, W; X = S, Se) monolayers show direct band gaps with the valence band maximum (VBM) and conduction band minimum (CBM) locating at the K point in the momentum space. In Figure 4.8, monolayer WS_2 is taken as an example to calculate the phononic dispersion in the first Brillouin zone. With three atoms in a unit cell, monolayer MX_2 should have nine phonon branches, of which the lowest three branches are the LA, TA, and out-of-plane acoustic modes (ZA), respectively. It can be seen that the LA and TA modes exhibit linear dispersion in the long wavelength limit, while the ZA mode shows an approximate quadratic dependence. The calculated longitudinal sound velocity is 4.3 cm/s for monolayer WS_2.

In Figure 4.9, the carrier–phonon scattering rates in WS_2 monolayer are calculated by Fermi's golden rule. The wave vector k of initial state is chosen along the K–Γ axis. Abrupt jumps are shown in the scattering rates, indicating the onset of

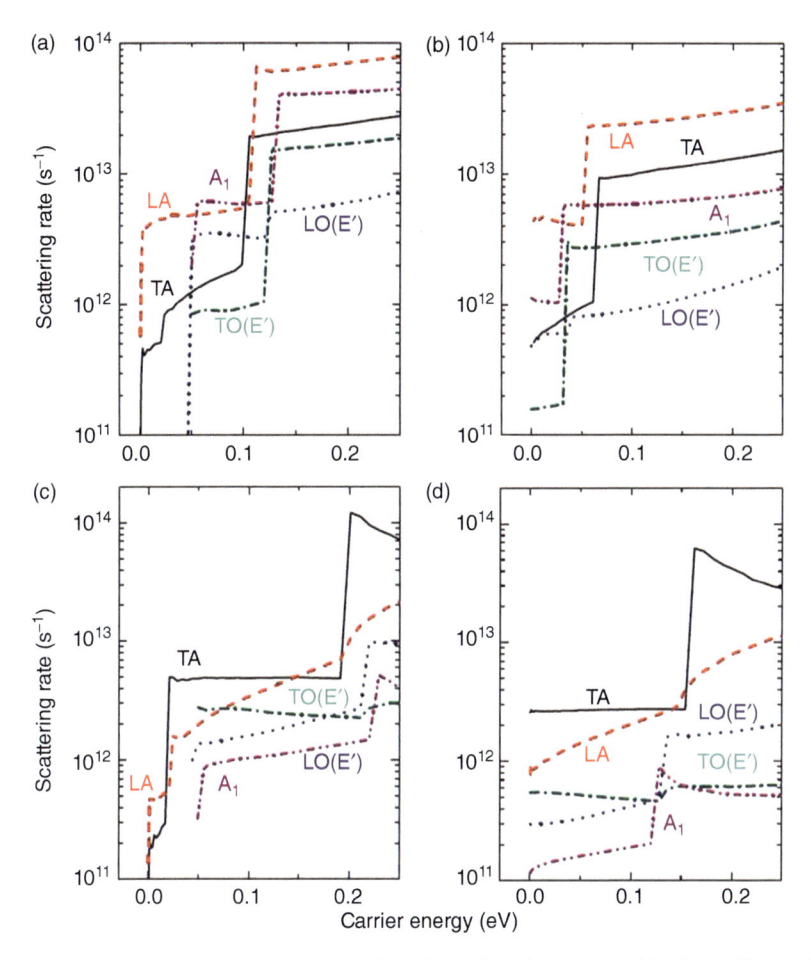

Figure 4.9 Intrinsic scattering rates of (a), (b) K-valley electrons, and (c, d) K-valley/peak holes in monolayer WS_2 via different phonon modes at room temperature. Panels (a, c) illustrate phonon emission, while (b, d) are for phonon absorption. Source: Reprinted with permission from Jin et al. [55]. Copyright 2014, American Physical Society.

intervalley transitions. Furthermore, it is found that the acoustic phonons play a dominant role in the scattering mechanism for both electrons and holes. What makes difference is that the LA mode shows the largest scattering rate for electrons, while it is the TA mode for holes. From Figure 4.9, it is clear that the optical branches exert weak influence on the scattering rates of low-energy carriers. This is attributed to the weak carrier–phonon interaction. It is found that MX_2 (M = Mo, W; X = S, Se) shows the similar scattering characteristics, because they have similar electronic structures at the VBM and the CBM at the K point in their band structure.

With the calculated scattering rates, the carrier mobility can be obtained by Monte Carlo simulation at different temperatures. As shown in Figure 4.10, the calculated mobility of electrons and holes are about 320 and 540 cm^2/V s,

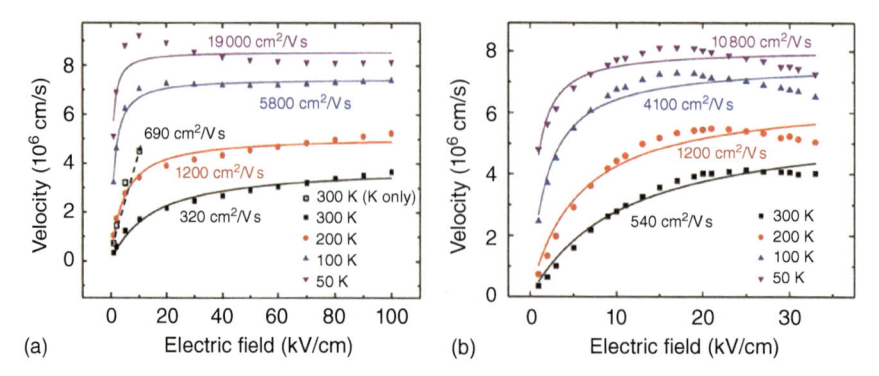

Figure 4.10 (a) Electron and (b) hole drift velocity versus electric field in monolayer WS$_2$ obtained by a full-band Monte Carlo simulation at different temperatures. Source: Reprinted with permission from Jin et al. [55]. Copyright 2014, American Physical Society.

respectively. Because of the smaller effective masses of electrons than that of holes at the K point in WS$_2$ monolayer, the higher mobility of holes is attributed to the low scattering rates. When the influence of K–Q intervalley scattering is corrected, the electron mobility at the K point is approximately 690 cm^2/V s. The results indicate that WS$_2$ monolayer has excellent transport properties, which shows good consistent with the experimental results.

Besides the phonon scattering, there are many other scattering mechanisms that limit the carrier mobility. As the Matthiessen rule, the total mobility can be written as

$$\frac{1}{\mu_{total}} = \frac{1}{\mu_{phonon}} + \frac{1}{\mu_{impurity}} + \frac{1}{\mu_{electron}} + \cdots$$

where the right items represent mobility limitation by the scattering of phonons, impurities, electrons, and so on. It is rather difficult to determine the contribution from each of these mechanisms, both experimentally and theoretically. Therefore, the total mobility is difficult to predict precisely. It is possible to search materials with high mobility by calculating their transport properties. More sophisticated calculations are necessary to determine the total mobility more precisely. However, these calculations are time consuming.

It has been demonstrated that the relativistic effect of spin–orbit coupling has a considerable effect on the electronic properties of TMDs. It is found that TMDs show sizable band splitting in the valence band. Taking the influence of the spin–orbit coupling on the carrier–phonon scattering into consideration, the calculated scattering matrix elements remain almost unchanged. Further research by Hinsche et al. has revealed a drastically different electron–phonon interaction in the spin–split valence band states at the K point in WS$_2$ monolayer. Their results have revealed a very long lifetime in the holes of the upper band. For the holes in the lower band, however, their lifetimes are significantly reduced by the fairly strong electron–phonon coupling. Because of this strong spin–orbit interaction in WS$_2$ monolayer, it can be expected that holes can be introduced in the upper band at the K point without changing the hole population in the lower band or the other branches. By such a scheme, high hole mobility can be obtained. This has been proved by the observed long lifetimes of a spin and valley-polarized holes in monolayer TMDs, making them candidates for topological superconductivity, which is different from the conventional electron–phonon-mediated superconductivity. Furthermore, Hinsche et al. have also demonstrated that the remarkable difference of electron–phonon coupling of the two spin–split branches even exists at the presence of a strongly interacting substrate.

4.2.3 Anisotropic Transport Properties of 2D Group-VA Semiconductors

2D Group-VA semiconductors exhibit superior carrier transport properties, making them good candidates for application in electronic and optoelectronic devices [56–61]. It is known that the carrier transport properties can be evaluated by calculating the carrier mobility. In theory, the carrier mobility of a 2D material can be calculated as the following expression [57]:

$$\mu_{2D} = \frac{2e\hbar^3 C_{2D}}{3k_B T |m^*|^2 E_1^2}$$

where $\hbar$, k_B, and T are the reduced Planck constant, Boltzmann constant, and temperature, respectively. m^* is the carrier-effective mass along the transport direction and C_{2D} is the elastic modulus. E_1 is the deformation potential constant of the VBM (hole) or the CBM (electron) along the transport direction. It is clear that the carrier mobility depends on three intrinsic properties of 2D materials including carrier-effective mass m^*, elastic modulus C_{2D}, and deformation potential constant E_1.

It is found that 2D phosphorene is a p-type semiconductor with a direct band gap at the Γ point. By using the above formula, Qiao et al. explored the transport properties of phosphorene by calculating its carrier-effective masses and mobilities [57]. Anisotropic mobilities have been demonstrated with different carrier mobilities along armchair and zigzag directions. Specifically, the electron mobility is 1100–1400 cm^2/V s along the armchair direction and ∼80 cm^2/V s along the zigzag direction. The hole mobility along the armchair direction is 640–700 cm^2/V s and up to 10 000 cm^2/V s along the zigzag direction, which is much higher than that of MoS_2. The extremely high hole mobility is mainly attributed to the rather small deformation potential at the VBM ($E_1 = 0.15 \pm 0.03$ eV). As shown in Figure 4.11, the highly atomic structure of phosphorene gives rise to the significant difference in effective mass, elastic modulus, and deformation potential constant along the armchair and zigzag directions, further leading to the carrier mobility anisotropy. These highly anisotropic transport properties of phosphorene make it distinguishing from other typical 2D materials, such as graphene, h-BN, MX_2 (M = Mo, W; X = S, Se), and so on.

The carrier transport properties of other 2D group-VA semiconductors, including arsenene, antimonene, and bismuthene, have also been studied by using the deformation potential method [62–65]. Carrier mobilities up to several thousand cm^2/V s are predicted in α- and β-arsenene, β-antimonene,

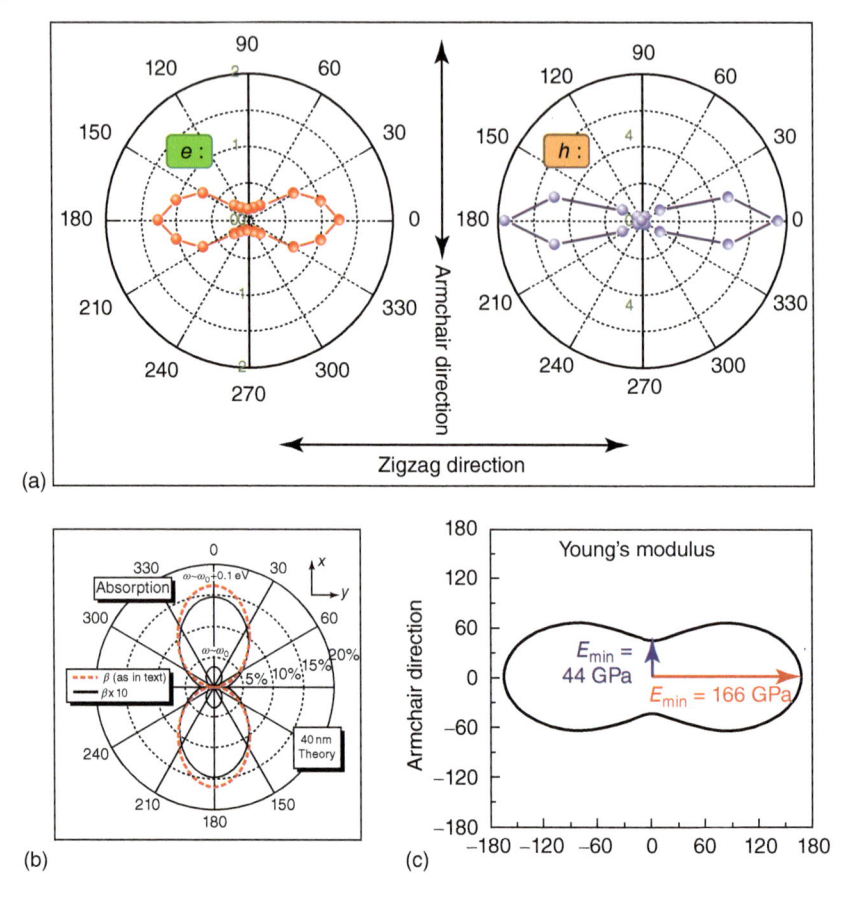

Figure 4.11 (a) Effective mass of electrons and holes according to spatial directions. Source: Reprinted with permission from Fei et al. 2014 [91]. Copyright 2014, American Chemical Society. (b) Polar representation of the absorption coefficient $A(\alpha)$ for a 40 nm intrinsic black phosphorus film for normal incident light with excitation energies at the band gap ω_0 and larger. α is the light polarization angle. $A(\alpha)$ is plotted for two values of interband coupling strengths. Source: Reprinted with permission from Fei et al. [91]. Copyright 2014, American Chemical Society. (c) Direction dependence of Young's modulus of phosphorene. Source: Reprinted with permission from Wei [92]. Copyright 2014, American Institute of Physics.

and β-bismuthene. Different from the case of phosphorene, spin–orbit coupling effects become more significant at the VBM of arsenene and bismuthene, which results in significant changes in the carrier-effective masses and the deformation potentials at the VBM. As a result, the hole mobilities of As and Bi monolayers are improved by 25% and 84%, respectively.

It is worthy to mention that 2D group IV–VI semiconductors, such as GeS, GeSe, SnS, and SnSe, have similar layered atomic structures with phosphorene where two kinds of atoms occupy the sites of phosphorus atoms in phosphorene [66–68]. These 2D materials have attracted much interest for their many remarkable properties. Similar to phosphorene, group-IV monochalcogenides are demonstrated to have anisotropic transport properties. Huang et al. revealed anisotropic characteristics in their atomic structures, elastic modulus, and carrier-effective masses [69]. Anisotropic behavior is also observed in the optical properties of group-IV monochalcogenides.

4.3 Contacts Between 2D Semiconductors and Metal Electrodes

4.3.1 Carrier Schottky Barriers at the Interfaces Between 2D Semiconductors and Metal Electrodes

Metal–semiconductor contact is a critical component in electronic and optoelectronic devices. Three major factors including tunnel barrier, Schottky barrier, and orbital overlap should be taken into consideration to evaluate the quality of the metal–semiconductor contacts. All the three factors play important roles in the electron injection efficiency. The tunnel barrier at metal–semiconductor interfaces can be obtained by calculating the electrostatic potential along the perpendicular direction of the interfaces. The orbital overlap can be determined by calculating the partial density of states, which is the density of states on specified atoms and orbitals. The Schottky barrier (SB) is another parameter of the most importance at metal–semiconductor interfaces. The SB height is defined as the energy difference between the Fermi level of metal electrode and the band edges (the VBM and CBM), which has a significant impact on device performance.

The formation of the SB at metal–semiconductor interfaces has been under debate for decades [70]. Assuming no interaction between the metal and the semiconductor, the SB height to an n-type semiconductor, Φ_e, is determined as

$$\Phi_e = W_{\text{metal}} - \chi_s$$

where W_{metal} is the work function of the metal and χ_s is the electron affinity of the semiconductor. Following this Schottky–Mott model, for a given semiconductor, the SB height should exhibit linear dependence on the work function of the contact metals. Therefore contact metals with low W_{metal} can be used to achieve small electron SB and those with high W_{metal} to achieve small hole SB. Ohmic contact can be achieved by using extremely high W_{metal} or low W_{metal} metals. This Schottky–Mott model shows discrepancy with experimental results. In fact, an extra voltage drop (V_{int}) exists at the metal–semiconductor interfaces because of the charge rearrangement at the interfaces. Therefore, as shown in Figure 4.12, the SB height can be written as [71]

$$\Phi_e = W_{\text{metal}} - \chi_s + eV_{\text{int}}$$

The weak van der Waals (vdW) interlayer coupling in 2D semiconductors makes them distinguish from conventional semiconductors, such as GaAs and ZnS. The absence of surface dangling bonds and gap states make the layered 2D semiconductors advantageous for nanoelectronics. When a 2D semiconductor makes contacts with metals, V_{int} can be extremely small because of slight charge transfer at the interfaces. The SB height can be lowered to improve the performance of 2D semiconductor-based electronics by using metals with different work functions.

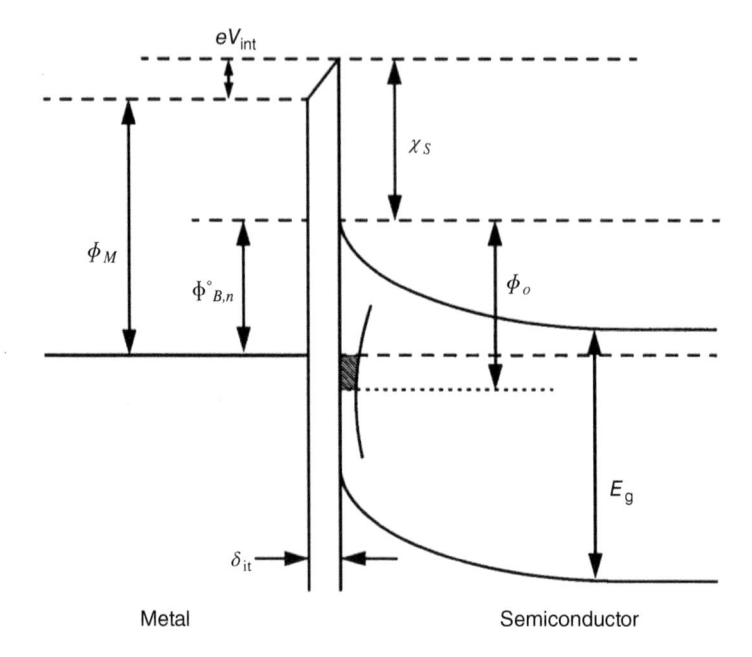

Figure 4.12 A schematic of the metal/semiconductor interface according to interface gap state models. Source: Reprinted with permission from Tung [71]. Copyright 2000, American Physical Society.

4.3.2 Partial Fermi Level Pinning and Tunability of Schottky Barrier at 2D Semiconductor–Metal Interfaces

The phenomena of Fermi level pinning (FLP) are usually induced by the defect states or interfacial states at the metal–semiconductor interface because of the presence of dangling bonds at the semiconductor surface. When a 2D semiconductor makes vdW contacts with metal electrodes, carrier transfer is normally impeded by the existing tunnel barrier at the contacting interfaces. Because of the weak vdW interaction between 2D materials and metals, a partial rather than complete FLP is predicted theoretically [72–74]. The partial FLP at interfaces between metals and 2D TMDs has been demonstrated, suggesting the tunability of the SB height by varying metal electrodes. For an n-type semiconductor, metal electrodes with sufficiently low work function are required to obtain a zero SB height. Bao et al. revealed a Quasi-Ohmic contact in MoS_2-based field effect transistors by using scandium as electrodes [75]. It is predicted that Cu(111) is a good candidate to achieve Ohmic contact to monolayer phosphorene [76].

Tuning the SB by varying the metal electrodes of different work functions underlies understanding the mechanism of the partial FLP at 2D semiconductor–metal systems. Bardeen et al. emphasized the role of surface states of the semiconductors in the FLP. Heine et al. ascribed the FLP to the metal-induced gap states whose density decays with the depth of semiconductors [77, 78]. For a 2D semiconductor, these two mechanisms are not applicable because of the absence of dangling bonds at the surfaces of 2D semiconductors [79].

Gong et al. have performed a systematic theoretical study of the partial FLP at interfaces of semiconductors and different metals with taking single-layer MoS$_2$ as an example [72]. The elemental metals are classified into two types with respect to the occupation of their d-orbitals, namely s-electron metals with fully occupied d-orbitals (Ag, Al, Au, Cd, Er, and Mg) and d-electron metals whose d-orbitals are not fully occupied (Hf, Ir, Pd, Pt, Sc, Ti, and Zr). Because of the existence of Mo d orbitals at the band edges of MoS$_2$, d-electron metals are revealed to have stronger interaction than s-electron metals. Partial density of states of MoS$_2$/metal interfaces revealed more orbital overlap between MoS$_2$ and d-electron metals.

To uncover the mechanism of partial FLP, Huang et al. have performed a systematical first-principles study of MoSe$_2$/metal contacts [80]. To determine the contact interaction, binding energies E_b between metals and MoSe$_2$ are calculated as

$$E_b = \frac{(E_{\text{total}} - E_{\text{metal}} - E_{\text{MoSe}_2})}{N_{\text{Mo}}}$$

where E_{total}, E_{metal}, and E_{MoSe_2} are the total energies of the combined systems, isolated metal electrodes, and isolated single-layer MoSe$_2$, respectively. N_{Mo} is the number of Mo atoms in corresponding systems. In Figure 4.13, Au and Pt are taken as examples of s-electron and d-electron metals to calculate the binding energy of MoSe$_2$ monolayer as a functional of interlayer distances. It is revealed that Pt obtains a significantly smaller interlayer distance and a stronger binding with MoSe$_2$ monolayer than Au. Similar results have been obtained by other works. To obtain further insight into the interaction between MoSe$_2$ and metals,

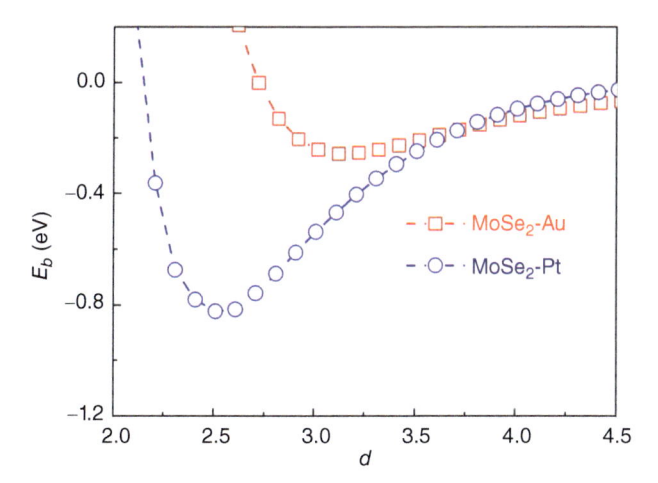

Figure 4.13 Binding energy E_b as a function of the separation d between MoSe$_2$ –Au(111) and MoSe$_2$–Pt(111) surface. Source: Reprinted with permission from Huang et al. [80]. Copyright 2017, American Chemical Society.

charge difference analysis is a useful method, which is calculated as

$$\Delta\rho(z) = \int \rho_{metal/MoSe_2} dxdy - \int \rho_{metal} dxdy - \int \rho_{MoSe_2} dxdy$$

Because there are no dangling bonds at the surface of $MoSe_2$ monolayer, physical adsorption rather than chemical bonding is expected between $MoSe_2$ and metals. As a result, a small amount of charge transfers between them.

The SB heights of electrons and holes can be determined by calculating the projected band structures of the metal/2D semiconductor systems. In Figure 4.14, projected band structures of metal/$MoSe_2$ systems are plotted. The electron Schottky barriers of the metal/$MoSe_2$ contacts are summarized in Figure 4.14. It is clear that both Φ_e and Φ_h vary linearly with the work function of metals (W_{metal}), with a slope of 0.67 (FLP coefficient). These results suggest a partial FLP in metal/$MoSe_2$ contacts. It is worth noting that s-electron metals and d-electron metals show almost the same FLP coefficient when contacting $MoSe_2$, regardless of their different interaction strengths.

Partial FLP suggests the tunability of the Schottky barriers in 2D semiconductor/metal systems. However, it is rather difficult to tune the Schottky barriers for 2D semiconductors by using different elemental metals of various work functions. Some other schemes have been proposed to overcome the FLP in 2D semiconductor/metal systems. It is explored by Liu et al. that by using 2D metals, such as T-MoS_2 and T-VS_2, rather than conventional metals, the Schottky barriers can be significantly tuned [81]. It is also proposed that the FLP can be reduced by inserting another layer of 2D metals between 2D semiconductors and metal electrodes [82–84].

4.3.3 Role of Defects in Enhanced Fermi Level Pinning in 2D Semiconductor/Metal Contacts

Although partial FLP in 2D semiconductor/metal contacts has been predicted by many theoretical works, it is rather difficult to reduce the carrier Schottky barriers (Φ_e and Φ_h) by varying the metal electrodes in laboratories. In this section, defects in the 2D semiconductor are ascribed to the enhanced FLP at 2D semiconductor/metal interfaces. It has been demonstrated that extra surface states induced by defects or dopants in semiconductors play an important role in the FLP at semiconductor/metal interfaces [77, 78, 85]. Theoretical works revealed the existence of defects in 2D semiconductors. Extra surface states induced by these defects can also result in enhanced FLP [79, 86–88]. The mechanism of the defect-enhanced FLP can be interpreted by Cowley's model [89]. The SB height in semiconductor/metal contacts is a linear combination of W_{metal} and a quantity φ_0, which is defined as the energy below which the surface states must be filled for charge neutrality at the semiconductor surface.

$$\Phi_e = \gamma(W_{metal} - \chi_s) + (1-\gamma)(E_g - \varphi_0)$$

where the parameter γ is given by

$$\gamma = \left(\frac{1 + (e^2 \delta_{it} D_{gs})}{\varepsilon_{it}} \right)^{-1}$$

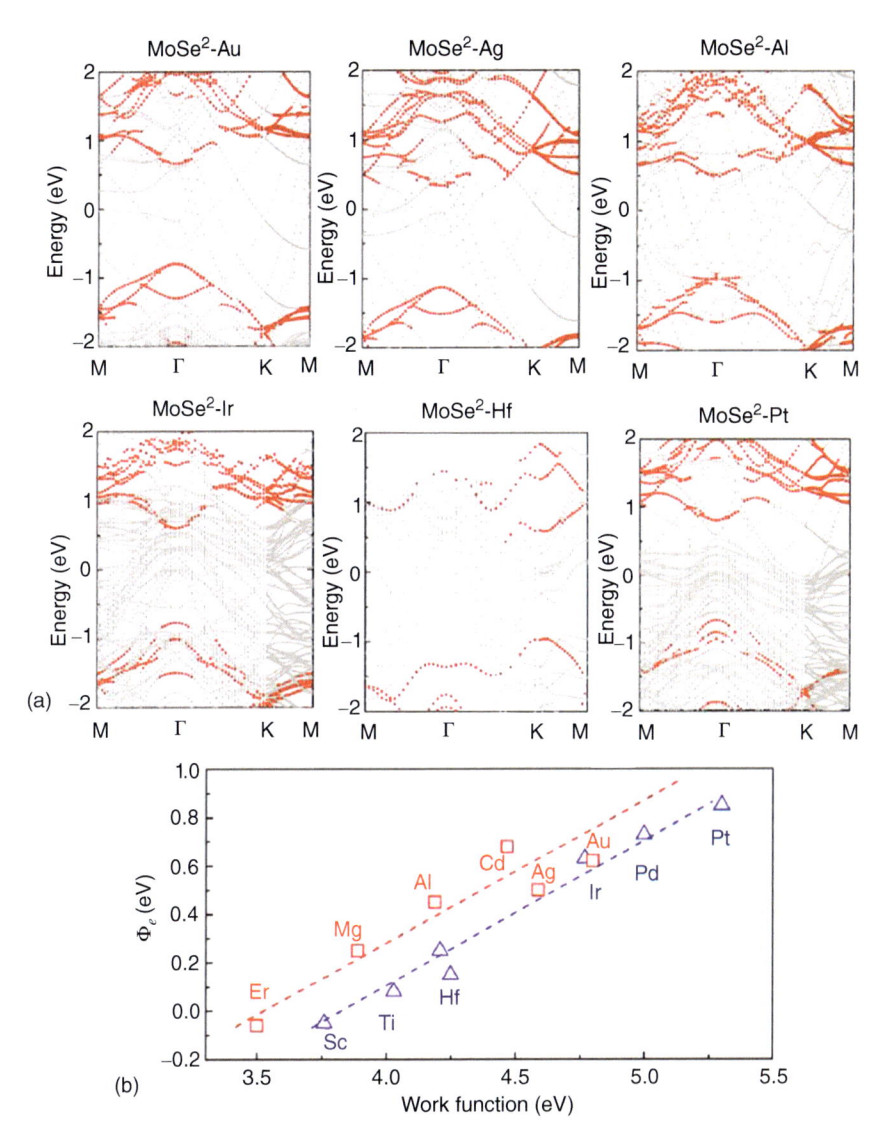

Figure 4.14 (a) Projected band structures of single-layer MoSe$_2$ contacting with several elemental metals. The Fermi level is set as zero. The bands dominated by MoSe$_2$ layer and metal atoms are plotted by red and gray dots, respectively. (b) Evolution of the Schottky barrier height of electrons (Φ_e) in MoSe$_2$/metal contacts as a function of the work function of MoSe$_2$/metal. The red squares and blue triangles present results of s-electron metals and d-electron metals, respectively. Source: Reprinted with permission from Huang et al. [80]. Copyright 2017, American Chemical Society.

For the case of semiconductors with high density of gap states (D_{gs}), a small γ indicates weak dependence of Φ_e on the metal work function. Strong FLP occurs at semiconductor/metal interfaces.

Huang et al. have performed a systematic study of the interfaces between several metals and perfect-, As_{Se}– (As substitutes Se), Br_{Se}–, and V_{Se}– (Se vacancy) $MoSe_2$ monolayers [90]. From the electronic band structures in Figure 4.15, it is found that As_{Se}, Br_{Se}, and V_{Se} introduce extra bands in the band structure of $MoSe_2$ monolayer. Specifically, the defect bands introduced by As_{Se} (Br_{Se}) are in

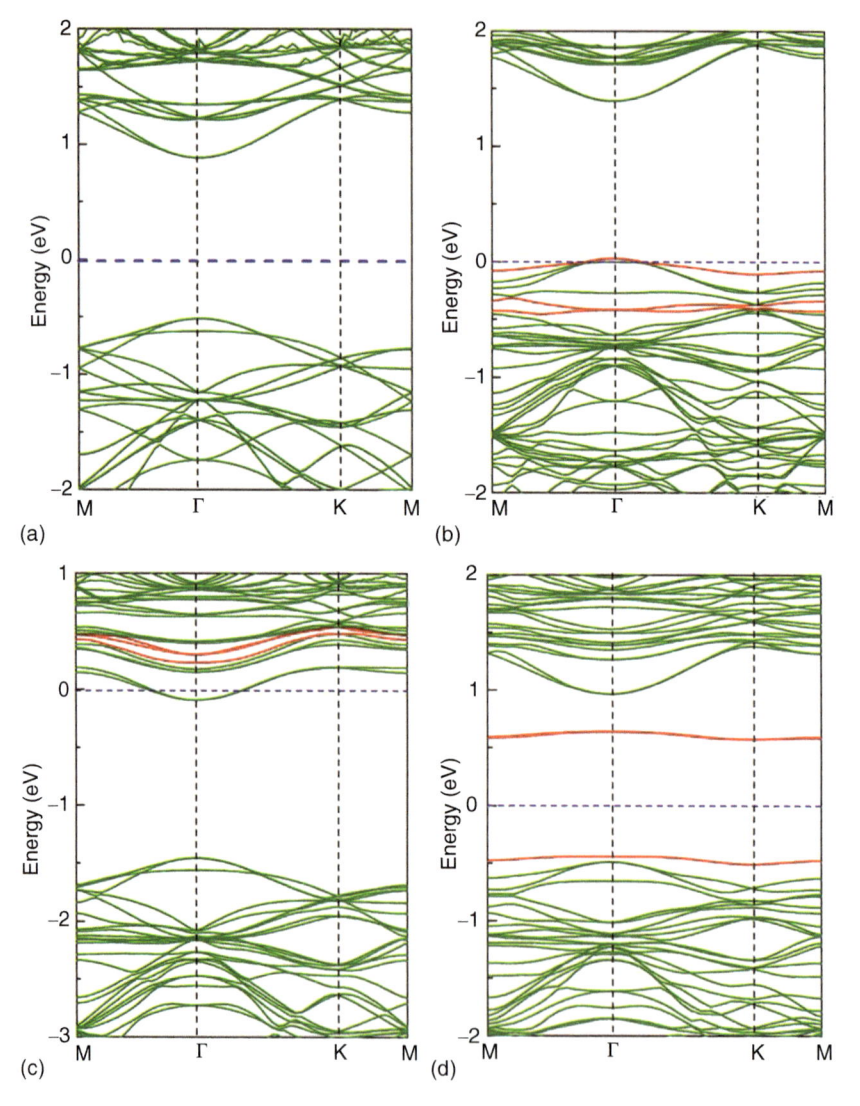

Figure 4.15 Band structures of (a) perfect-$MoSe_2$, (b) As_{Se}–$MoSe_2$, (c) Br_{Se}–$MoSe_2$, and (d) V_{Se}–$MoSe_2$. Some important impurity bands and defect bands are plotted by red lines. The Fermi level is taken as a reference. Source: Reprinted with permission from Huang et al. [90]. Copyright 2017, American Physical Society.

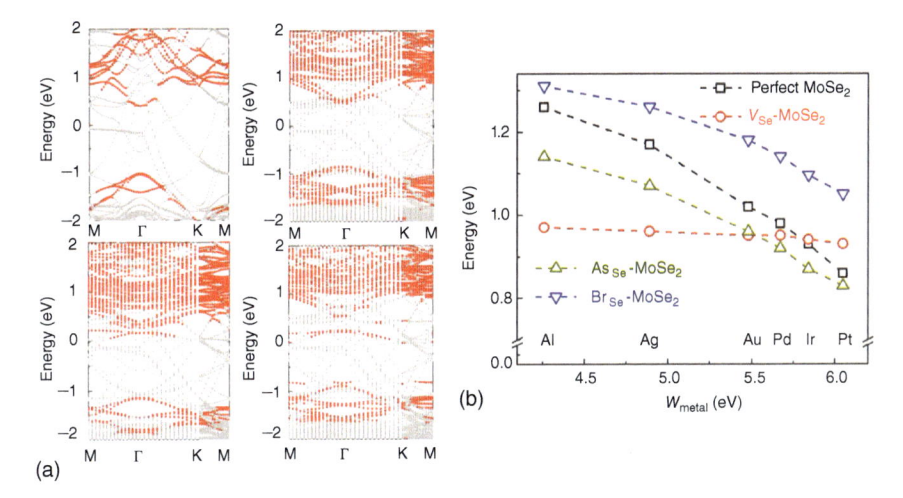

Figure 4.16 Projected band structures of (a) perfect-MoSe$_2$/Au, As$_{Se}$–MoSe$_2$/Au, Br$_{Se}$–MoSe$_2$/Au, and V$_{Se}$–MoSe$_2$/Au. Bands dominated by corresponding MoSe$_2$ layer and Au electrode are plotted by red and gray dots, respectively. The Fermi level is taken as a reference. (b) Variation of electron and hole Schottky barrier height as functions of the work function of metal electrodes. The valence band maximum of MoSe$_2$ is set to zero. Source: Reprinted with permission from Huang et al. [90]. Copyright 2017, American Physical Society.

the valence (conduction) bands because of one less (more) valence electron per As (Br) atom than Se atom. Compared with As$_{Se}$ and Br$_{Se}$, V$_{Se}$ produces extra defect states in the band gap of MoSe$_2$ monolayer with a much higher density. It is revealed that these defect states play a crucial role in the FLP of metal/MoSe$_2$ interfaces. When As$_{Se}$–, Br$_{Se}$–, and V$_{Se}$–MoSe$_2$ make contacts with Au, significant changes are observed in the carrier Schottky barrier (Φ_e and Φ_h) compared with perfect-MoSe$_2$/Au contact. As shown in Figure 4.16, perfect-, As$_{Se}$–, and Br$_{Se}$–MoSe$_2$/metal contacts exhibit partial FLP with the largest FLP coefficient in perfect-MoSe$_2$/metal systems. These results indicate that the interfacial states introduced by As$_{Se}$ and Br$_{Se}$ give rise to enhanced FLP. For V$_{Se}$–MoSe$_2$/metal contacts, Φ_e and Φ_h show little change with W_{metal}, suggesting a complete FLP. The inevitable vacancies at the interfaces may be the reason for strong FLP in experiments.

References

1 Novoselov, K.S., Geim, A.K., Morozov, S.V. et al. (2004). *Science* 306: 666.
2 Castro Neto, A.H., Guinea, F., Peres, N.M.R. et al. (2009). *Rev. Mod. Phys.* 81: 109.
3 Li, X., Wang, X., Zhang, L. et al. (2008). *Science* 319: 1229.
4 Tombros, N., Jozsa, C., Popinciuc, M. et al. (2007). *Nature (London)* 448: 571.
5 Xiang, H., Kan, E.J., Wei, S.-H. et al. (2009). *Nano Lett.* 9: 4025.
6 Topsakal, M., Bagci, V.M.K., and Ciraci, S. (2010). *Phys. Rev. B* 81: 205437.
7 Hu, C.H., Wu, S.Q., Wen, Y.H. et al. (2010). *J. Phys. Chem. C* 114: 19673.

8 Son, Y.-W., Cohen, M.L., and Louie, S.G. (2006). *Phys. Rev. Lett.* 97: 216803.
9 Pisani, L., Chan, J.A., Montanari, B., and Harrison, N.M. (2007). *Phys. Rev. B* 75: 064418.
10 Li, Z., Qian, H., Wu, J. et al. (2008). *Phys. Rev. Lett.* 100: 206802.
11 Yan, Q., Huang, B., Yu, J. et al. (2007). *Nano Lett.* 7: 1469.
12 Kang, J., Wu, F., and Li, J. (2011). *Appl. Phys. Lett.* 98: 083109.
13 Jiang, J., Lu, W., and Bernholc, J. (2008). *Phys. Rev. Lett.* 101: 246803.
14 Kobayashi, K., Fukui, K., Enoki, T., and Kusakaba, K. (2006). *Phys. Rev. B* 73: 125415.
15 Kan, E., Wu, X., Li, Z. et al. (2008). *J. Chem. Phys.* 129: 084712.
16 Dutta, S., Manna, A., and Pati, S. (2009). *Phys. Rev. Lett.* 102: 096601.
17 Ozaki, T., Nishio, K., Weng, H., and Kino, H. (2010). *Phys. Rev. B* 81: 075422.
18 Deng, X.Q., Zhang, Z.H., Tang, G.P. et al. (2014). *Carbon* 66: 646–653.
19 Buttiker, M. and Landauer, R. (1985). *Phys. Rev. B* 31: 6207.
20 Kim, W. and Kim, K. (2008). *Nat. Nanotechnol.* 3: 40.
21 Kan, E., Li, Z., Yang, J., and Hou, J. (2008). *J. Am. Chem. Soc.* 130: 4224–4225.
22 Son, Y., Cohen, M., and Louie, S. (2006). *Nature (London)* 444: 347.
23 Tang, G., Zhou, J., Zhang, Z. et al. (2013). *Carbon*: 94–101.
24 Guo, Y., Jiang, K., Xu, B. et al. (2012). *J. Phys. Chem. C* 116: 13837.
25 Cranford, S.W. and Buehler, M.J. (2011). *Carbon* 49: 4111.
26 Pan, L.D., Zhang, L.Z., Song, B.Q. et al. (2011). *Appl. Phys. Lett.* 98: 173102.
27 Jiao, Y., Du, A., Hankel, M. et al. (2011). *Chem. Commun.* 47: 11843.
28 Kang, J., Li, J., Wu, F. et al. (2011). *J. Phys. Chem. C* 115: 20466.
29 Long, M., Tang, L., Wang, D. et al. (2011). *ACS Nano* 5: 2593.
30 Narita, N., Nagai, S., Suzuki, S., and Nakao, K. (1998). *Phys. Rev. B* 58: 11009.
31 Malko, D., Neiss, C., Vines, F., and Gorling, A. (2012). *Phys. Rev. Lett.* 108: 086804.
32 Malko, D., Neiss, C., and Gorling, A. (2012). *Phys. Rev. B* 86: 045443.
33 Yue, Q., Chang, S., Kang, J. et al. (2012). *J. Chem. Phys.* 136: 244702.
34 Kim, W.Y., Choi, Y.C., Min, S.K. et al. (2009). *Chem. Soc. Rev.* 38: 2319.
35 Yue, Q., Chang, S., Tan, J. et al. (2012). *Phys. Rev. B* 86: 235448.
36 Cahangirov, S., Topsakal, M., Akturk, E. et al. (2009). *Phys. Rev. Lett.* 102: 236804.
37 Ding, Y. and Ni, J. (2009). *Appl. Phys. Lett.* 95: 083115.
38 Wang, Y., Zheng, J., Ni, Z. et al. (2012). *Nano: Brief Rep. Rev.* 7: 1250037.
39 Kang, J., Wu, F., and Li, J. (2012). *Appl. Phys. Lett.* 100: 233122.
40 Borysenko, K.M., Mullen, J.T., Barry, E.A. et al. (2010). *Phys. Rev. B* 81: 121412(R).
41 Baroni, S., de Gironcoli, S., and Dal Corso, A. (2001). *Rev. Mod. Phys.* 73: 515.
42 Novoselov, K.S., Jiang, D., Schedin, F. et al. (2005). *Proc. Natl. Acad. Sci. U.S.A.* 102: 10451.
43 Radisavljevic, B., Radenovic, A., Brivio, J. et al. (2011). *Nat. Nanotechnol.* 6: 147.
44 Lee, C., Yan, H., Brus, L.E. et al. (2010). *ACS Nano* 4: 2695.
45 Mak, K.F., Lee, C., Hone, J. et al. (2010). *Phys. Rev. Lett.* 105: 136805.

References

46 Korn, T., Heydrich, S., Hirmer, M. et al. (2011). *Appl. Phys. Lett.* 99: 102109.
47 Splendiani, A., Sun, L., Zhang, Y. et al. (2010). *Nano Lett.* 10: 1271.
48 Jariwala, D., Sangwan, V.K., Lauhon, L.J. et al. (2014). *ACS Nano* 8: 1102.
49 Radisavljevic, B. and Kis, A. (2013). *Nat. Mater.* 12: 815.
50 Fang, H., Chuang, S., Chang, T.C. et al. (2012). *Nano Lett.* 12: 3788.
51 Liu, W., Kang, J., Sarkar, D. et al. (2013). *Nano Lett.* 13: 1983.
52 Fuhrer, M.S. and Hone, J. (2013). *Nat. Nanotechnol.* 8: 146.
53 Radisavljevic, B. and Kis, A. (2013). *Nat. Nanotechnol.* 8: 147.
54 Hwang, E.H. and Das Sarma, S. (2008). *Phys. Rev. B* 77: 235437.
55 Jin, Z., Li, X., Mullen, J.T., and Kim, K.W. (2014). *Phys. Rev. B* 90: 045422.
56 Li, L., Yu, Y., Ye, G.J. et al. (2014). *Nat. Nanotechnol.* 9: 372–377.
57 Qiao, J., Kong, X., Hu, Z.-X. et al. (2014). *Nat. Commun.* 5: 4475.
58 Buscema, M., Groenendijk, D.J., Blanter, S.I. et al. (2014). *Nano Lett.* 14: 3347–3352.
59 Fei, R. and Yang, L. (2014). *Nano Lett.* 14: 2884–2889.
60 Han, X., Morgan Stewart, H., Shevlin, S.A. et al. (2014). *Nano Lett.* 14: 4607–4614.
61 Huang, L., Huo, N., Li, Y. et al. (2015). *J. Phys. Chem. Lett.* 6: 2483.
62 Zhang, S., Xie, M., Li, F. et al. (2016). *Angew. Chem. Int. Ed. Engl.* 128: 1698–1701.
63 Pizzi, G., Gibertini, M., Dib, E. et al. (2016). *Nat. Commun.* 7: 12585.
64 Wang, Y., Huang, P., Ye, M. et al. (2017). *Chem. Mater.* 29: 2191–2201.
65 Zhang, S., Guo, S., Chen, Z. et al. (2018). *Chem. Soc. Rev.* 47: 982–1021.
66 Zhao, L.-D., Lo, S.-H., Zhang, Y. et al. (2014). *Nature* 508: 373.
67 Hu, Y., Zhang, S., Sun, S. et al. (2015). *Appl. Phys. Lett.* 107: 122107.
68 Shi, G. and Kioupakis, E. (2015). *Nano Lett.* 15: 6926.
69 Huang, L., Wu, F., and Li, J. (2016). *J. Chem. Phys.* 144: 114708.
70 Tung, R.T. (1993). *J. Vac. Sci. Technol. B* 11: 1546.
71 Tung, R.T. (2000). *Phys. Rev. Lett.* 84: 6078.
72 Gong, C., Colombo, L., Wallace, R.M., and Cho, K. (2014). *Nano Lett.* 14: 1714–1720.
73 Chen, W., Santos, E.J., Zhu, W. et al. (2013). *Nano Lett.* 13: 509–514.
74 Popov, I., Seifert, G., and Tomanek, D. (2012). *Phys. Rev. Lett.* 108: 156802.
75 Bao, W., Cai, X., Kim, D. et al. (2013). *Appl. Phys. Lett.* 102: 042104.
76 Gong, K., Zhang, L., Ji, W., and Guo, H. (2014). *Phys. Rev. B* 90: 125441.
77 Heine, V. (1965). *Phys. Rev.* 138: 1689.
78 Louie, S.G. and Cohen, M.L. (1975). *Phys. Rev. Lett.* 35: 866–869.
79 Zhou, W., Zou, X., Najmaei, S. et al. (2013). *Nano Lett.* 13: 2615–2622.
80 Huang, L., Li, B., Zhong, M. et al. (2017). *J. Phys. Chem. C* 121: 9305.
81 Liu, Y., Stradins, P., and Wei, S.-H. (2016). *Sci. Adv.* 2: e1600069.
82 Musso, T., Kumar, P.V., Foster, A.S., and Grossman, J.C. (2014). *ACS Nano* 8: 11432–11439.
83 McDonnell, S., Azcatl, A., Addou, R. et al. (2014). *ACS Nano* 8: 6265–6272.
84 Chuang, S., Battaglia, C., Azcatl, A. et al. (2014). *Nano Lett.* 14: 1337–1342.
85 Hasegawa, H. and Sawada, T. (1983). *Thin Solid Films* 103: 119–140.

86 Himpsel, F.J., Hollinger, G., and Pollak, R.A. (1983). *Phys. Rev. B* 28: 7014.
87 Nishimura, T., Kita, K., and Toriumi, A. (2007). *Appl. Phys. Lett.* 91: 123123.
88 Segev, D. and Van de Walle, C.G. (2007). *Europhys. Lett.* 76: 305.
89 Cowley, A.M. and Sze, S.M. (1965). *J. Appl. Phys.* 36: 3212.
90 Huang, L., Tao, L., Gong, K. et al. (2017). *Phys. Rev. B* 96: 205303.
91 Fei, R., Faghaninia, A., Soklaski, R. et al. (2014). *Nano Lett.* 14: 6393–6399.
92 Wei, Q. and Peng, X. (2014). *Appl. Phys. Lett.* 104: 251915.

5

Preparation and Properties of 2D Semiconductors

5.1 Preparation Methods

5.1.1 Mechanical Exfoliation

So far, the mechanical exfoliation, liquid-phase exfoliation, and chemical vapor deposition (CVD) are the three main methods to synthesize 2D semiconductors. Monolayer and few-layer 2D semiconductors can be obtained by these methods. However, among them, the mechanical exfoliation can yield the high-quality atomically thin crystals, which are ideal for studying the fundamental physical and chemical properties of 2D crystals. This method is considered as a simple, fast, and cost-effective technique to obtain the atomically thin crystals of 2D semiconductors. However, this method also has some disadvantages such as low yield and poor controllability. Therefore, this method is very suitable for basic research in the laboratory, but it is ineffective for large-scale practical applications [1–3].

The standard mechanical exfoliation involves three main steps [4]. First, using tweezers to peel the 2D crystals out of a flat, fresh surface and pressing this flat, fresh crystal on the sticky side of a piece of tape (3M Scotchs tape). Then, another piece of tape attaches to this tape with the flat, fresh crystal. These two pieces of tapes are quickly pulled apart and adhering the same size of crystals on each of them. Sticking the tape with crystals on the clean substrate and gently touching it by band or the soft object. Finally, the tape is peeled from the substrate to leave the 2D crystals on the substrate. The tackiness of the tape, the experimental temperature and humidity, and the substrate directly affect the size of the final peeled 2D crystals. Usually, SiO_2/Si is generally selected as the substrate for the mechanical exfoliation of 2D materials. This is mainly due to the compatibility of the SiO_2/Si substrate with the integrated devices, which is beneficial to the study of the optoelectronic/electronic properties of 2D material-based micro/nanodevices. The substrate must be cleaned before using it. To remove the surface contaminants, the SiO_2/Si substrates were cleaned by the typical RCA cleaning method and then in acetone and isopropanol in an ultrasonic bath for 15 minutes and finally blow-dried with nitrogen. For other applications, other materials such as sapphire, mica are also selected as the substrates for the mechanical exfoliation of 2D materials. Figure 5.1 shows the optical microscope and atomic force microscope (AFM) images of a typical MoS_2 sample exfoliated from its bulk crystals [4].

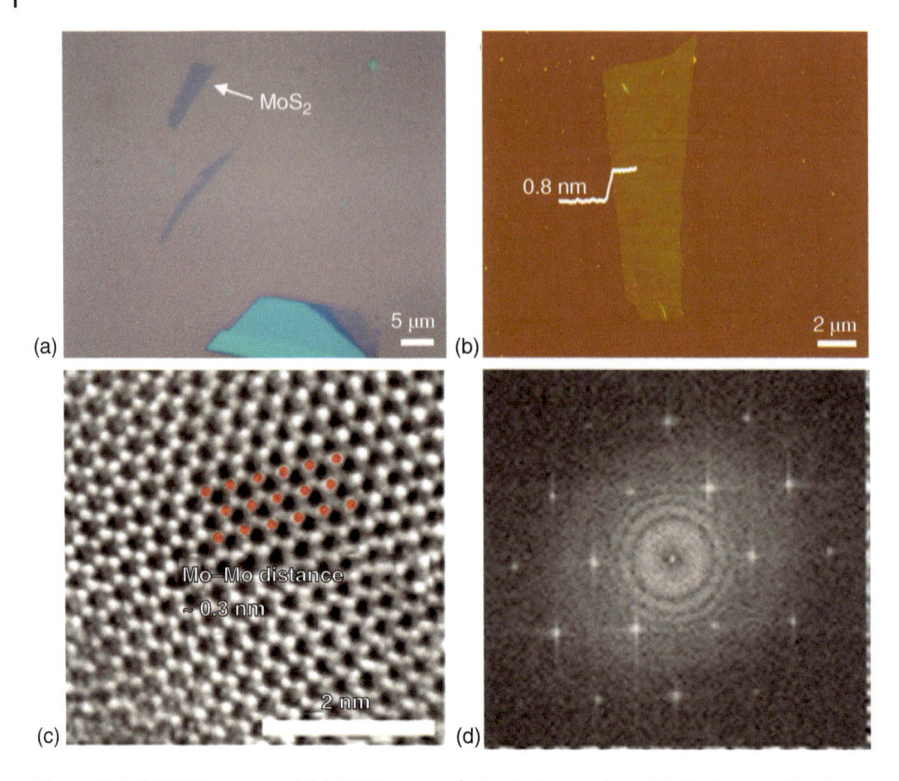

Figure 5.1 (a) OM image and (b) AFM image of a typical monolayer MoS_2. Source: Reprinted with permission from Yin et al. [4]. Copyright 2012, American Chemical Society. (c) HRTEM and (d) fast Fourier transform (FFT) image of a typical monolayer MoS_2. The monolayer MoS_2 samples exfoliated from its bulk crystal. Source: Reprinted with permission from Najmaei et al. [5]. Copyright 2012, American Institute of Physics.

According to the high-resolution transmission electron microscopy (HRTEM) results, we can know that this sample has a cleanliness surface and good quality [5]. Other 2D materials such as WS_2, WSe_2, WTe_2, $MoSe_2$, and $MoTe_2$ also have been obtained from their bulk crystals by this technique [6–10].

The traditional mechanical exfoliation is limited in producing large surface area crystals with the thickness of a few atom layers. To increase the yield of high-quality crystals, many alternative methods have been developed for the exfoliation of 2D materials. Huang et al. reported a modified approach for exfoliating thin flakes from 2D crystals [11]. Before exfoliating few-layer flakes from bulk crystal, the SiO_2/Si substrate is cleaned in acetone, 2-propanol, and deionized (DI) water by an ultrasonic bath. Then, the substrate was cleaned by oxygen plasma to remove the surface contaminants. Following the plasma cleaning step, the tape with 2D crystals is attached on the clean substrate. After that, the substrate with the attached tape is annealed for two to five minutes at about 100 °C in air. Finally, the adhesive tape is removed when the sample cooled to room temperature. These two modified process steps enhance and homogenize the adhesion force between crystals and substrate, which are very important for the yield of high-quality crystals with large lateral size. Figure 5.2 shows the

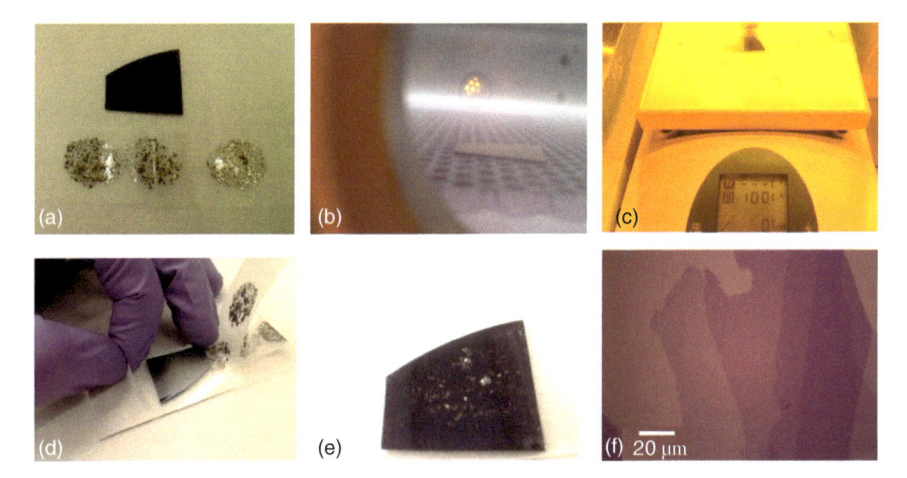

Figure 5.2 (a) Substrate and prepared crystals on tapes. (b–e) Huang's modified exfoliation process for layered crystals. (f) Optical image of the exfoliated crystal. Source: Reprinted with permission from Huang et al. [11]. Copyright 2015, American Chemical Society.

crystals obtained by this method. According to statistical results, the yield and area of the few-layer crystals increased more than 50 times compared to the traditional exfoliation methods. Meanwhile, the crystals show high-quality and excellent electronic properties. This modified exfoliation method is considered as an effective way to produce large-area, high-quality thin flakes of 2D materials. In addition, Desai et al. developed an Au film-assisted exfoliation method, and the detailed process shown in Figure 5.3 [12]. First, gold film is evaporated onto 2D crystals and the evaporated gold atoms interact with the topmost layer. The interaction between gold and crystals is stronger than the van der Waals interactions in 2D crystals. After that, a thermal release tape is used to stick the gold film with thin crystals onto the SiO_2/Si substrate. Then, the thermal release tape is placed on a hot plate and the gold film will release from the thermal release tape on the substrate. Finally, we use potassium iodide and iodine (KI/I_2) to etch the gold film and the thin crystals will leave on the substrate. This method can also prepare large lateral size thin crystals with high quality.

5.1.2 Liquid-Phase Exfoliation

Liquid-phase exfoliation has been used to prepare a variety of 2D materials and is considered as an effective way to obtain a large quality of thin layer 2D materials. In such process, three major steps are usually needed: first, the bulk 2D crystal is dispersed in a suitable solvent; then, the bulk 2D crystal exfoliated into thin flakes with the aid of equipment; finally, the thin flakes separated from the solvent by the centrifugation. The type of solvent used in exfoliation directly affects the quality of the obtained samples. To stop the reaggregation of exfoliated few-layer materials, the selected solvent must have comparable surface tension with the surface energy of 2D materials. A wide variety of solvents, such as water, dimethylformamide (DMF), N-methyl-2-pyrrolidone (NMP),

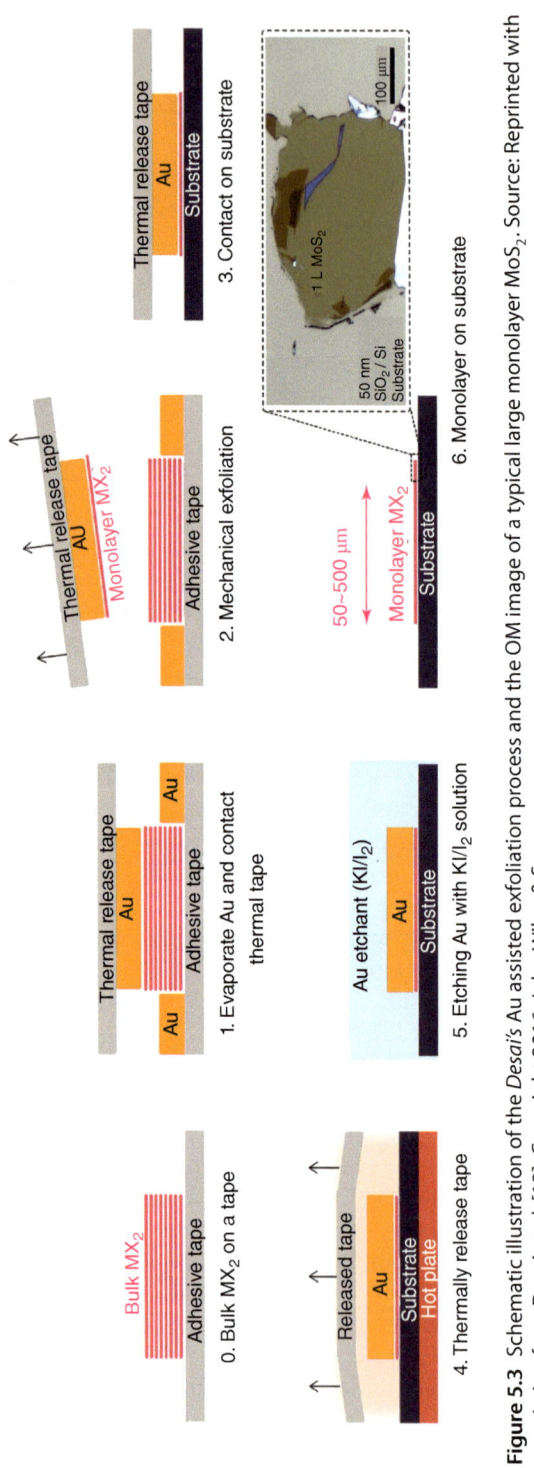

Figure 5.3 Schematic illustration of the *Desai's* Au assisted exfoliation process and the OM image of a typical large monolayer MoS_2. Source: Reprinted with permission from Desai et al. [12]. Copyright 2016, John Wiley & Sons.

dimethylsulfoxide (DMSO), *N*-viny-pyrrolidone (NVP), indole-3-propionic acid (IPA), *N*-dodecyl-pyrrolidinone, and cyclohexyl-pyrrolidinone, have been studied in detail to exfoliate 2D crystals [13, 14]. The ultrasonic energy, ultrasonic time, and ultrasonic temperature affect the yield and thickness of thin flakes. Improvements in concentrations of graphene dispersions have been achieved by using drastically longer sonication times (~2 mg/ml after 500 hours), since the first successful exfoliation of graphene in organic solvents in 2008 [15]. In addition to the energy consumption, the crystal size is severely reduced with the increase of sonication time. Studies also show that the long sonication time also affects the quality of the obtained 2D thin flakes. Some atomic or point defects are formed in the graphene after a long sonication process. In addition, the centrifugation time and speed have direct effect on the lateral size and thickness of exfoliated flakes. This technique has got great success in the preparation of single-layer or multilayer nanosheets of grapheme, transition metal dichalcogenides (TMDCs), black phosphorus, InSe, SnS_2, $SnSe_2$, and so on [15–19]. The samples prepared by this method have relatively small lateral size, which are highly suitable for the applications in chemical and energy fields such as catalysis, photovoltaic, chemical battery, and electrochemical energy storage but not suitable for the applications in nanoelectronic and optoelectronic devices.

The interlayers of 2D materials are bonded by the weak van der Waals force and have large distance in interlayers. Therefore, the organic or ionic species can intercalate into interlayers of bulk crystals easily. This process increases the interlayer spacing and weakens the interlayer adhesion [20]. Typically, few-layer MoS_2 flakes have been successfully exfoliated from its bulk crystal with *n*-butyl lithium dissolved in hexane as the intercalation agent. Firstly, the bulk MoS_2 crystals dispersed in *n*-butyl lithium solutions for more than one day to allow lithium ions to intercalate into interlayers and to form the Li_xMoS_2 compound. Then, the compound Li_xMoS_2 reacts with the lithium to release hydrogen gas by putting them into water. Finally, the thin flakes are separated by the centrifugation process. It is very effective for the preparation of high yield of few-layer MoS_2 flakes, but some serious problems also exist. According to the studies, the exfoliated monolayer MoS_2 has an octahedral structure ($1T$-MoS_2). We know that the natural bulk MoS_2 crystals have a trigonal prismatic structure ($2H$-MoS_2). This process results in the changes of structure and electronic properties of crystals. To obtain few-layer $2H$-MoS_2, other treatment such as annealing is must needed. Meanwhile, this annealing process is carried out at high temperature and long experiment period. In order to address these problems, Zeng et al. improved the exfoliation process by using an electrochemical cell with a lithium foil anode and transition metal chalcogenides (TMDs) cathode [21]. Experimental results show that a large number of few-layer TMDs flakes with microscale in lateral size are obtained after several hours, which are much faster than *n*-butyl lithium exfoliation. Meanwhile, in this process, the degree of lithium insertion can be monitored and controlled by controlling the galvanostatic discharge in the electrochemical cell.

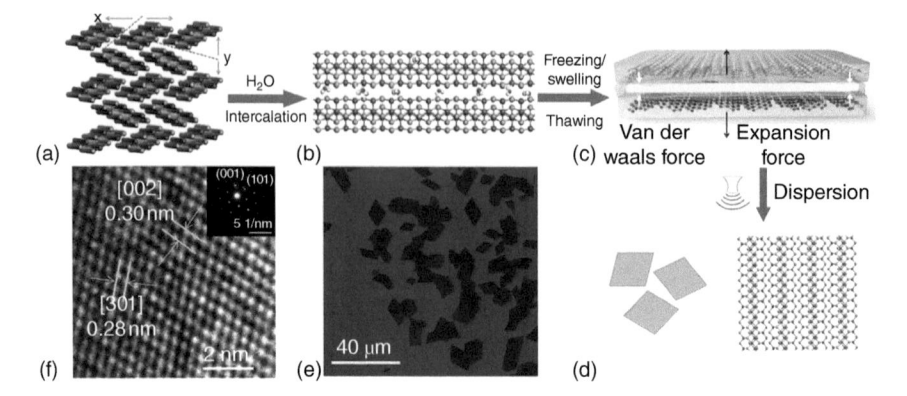

Figure 5.4 (a–d) Schematic demonstration of the gentle water freezing–thawing exfoliation strategy of 2D crystal. (e) SEM image and (f) HRTEM image of the prepared Sb_2Se_3 crystal. The inset image in (f) is the corresponding selected area electron diffraction (SAED) pattern. Source: Reprinted with permission from Song et al. [22]. Copyright 2017, John Wiley & Sons.

Jiang Tang group developed a gentle water freeze–thaw approach to prepare Sb_2S_3, Sb_2Se_3, Bi_2S_3, and $Sb_2S_xSe_{3-x}$ thin nanosheets [22]. First, the commercial Sb_2Se_3 Sb_2S_3, Bi_2S_3, and $Sb_2S_xSe_{3-x}$ alloy powders were dispersed into H_2O with vigorous stirring for 24 hours. Subsequently, the mixture was kept in a refrigerator for water intercalation and freezing for a few hours. After that, the frozen mixture was ultrasonically dispersed into water for 40 minutes at room temperature, and this process was repeated six times. Finally, after filtering, centrifugation, and ultrasonic treatment, the ultrathin flake samples are obtained. The detailed process is demonstrated in Figure 5.4. The high quality of the samples is characterized by the Raman and HRTEM results. The photodetection study also shows that they have high responsivity and anisotropic in-plane transport.

Recently, Xiangfeng Duan group reported a general method to prepare highly uniform, solution-processable, phase-pure 2D nanosheets [23]. The process involves three steps (Figure 5.5): the quaternary ammonium molecules intercalate into 2D crystals by electrochemical technology, mild sonication, and centrifugation process. The tetraheptylammonium bromide (THAB) molecules inserted into the MoS_2 crystal by electrochemical technology, and THA^+ cations inserted into the MoS_2 layer were driven by the negative electrochemical potential. As a result, this process causes the volume expansion of the MoS_2 crystal. Then, the processed MoS_2 crystals were dispersed into the polyvinylpyrrolidone solution in dimethylformamide (PVP/DMF) and sonicated and the MoS_2 crystals were processed for several minutes to produce thin MoS_2 nanosheets. Finally, the specific concentration of MoS_2 nanosheets was obtained by tuning the centrifugation. This method can prepare high-quality thin crystals, and the thin crystals also show excellent electronic properties. Meanwhile, this method has been used to prepare a wide range of 2D materials, including TMDCs, Bi_2Se_3, In_2Se_3, Sb_2Te_3, and black phosphorus, and considered as a potential method to the scalable production of high-quality thin nanosheets for large-area electronics, optoelectronics, and thermoelectrics.

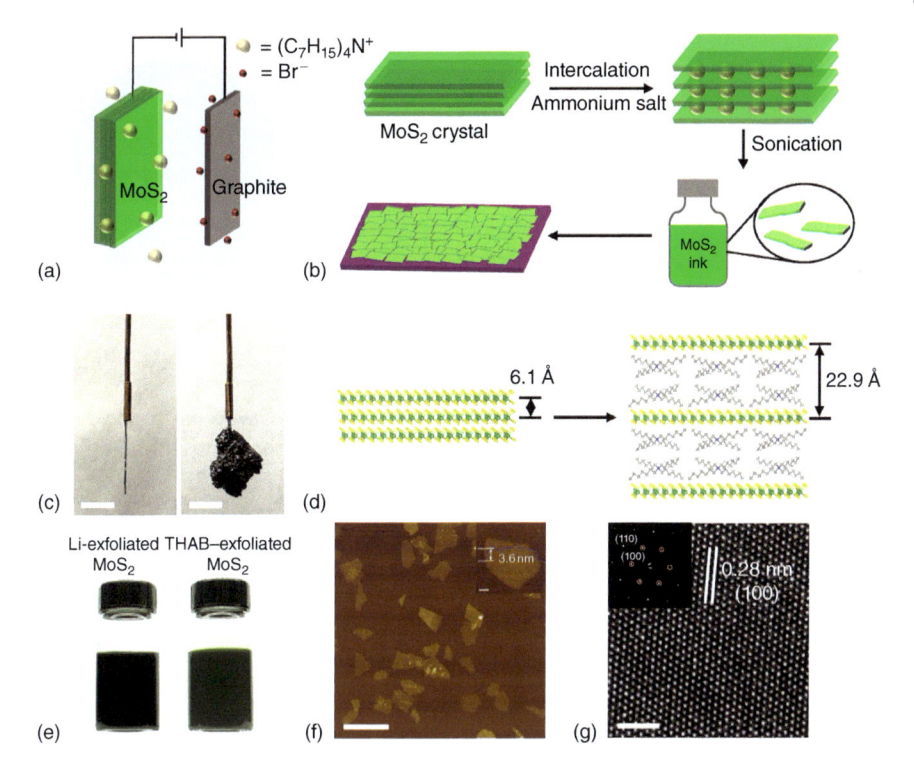

Figure 5.5 (a–d) Schematic illustration of the electrochemical intercalation of MoS_2 with molecular. (e) Photograph of THAB-exfoliated MoS_2 crystals in isopropanol and Li-exfoliated MoS_2 crystals in water. (f) AFM image of the prepared few-layer MoS_2 crystals. Scale bar is 2 μm. (g) The HRTEM image of a typical MoS_2 crystal. The inset image in (g) is the corresponding SAED pattern. Scale bar is 2 nm. Source: Reprinted with permission from Lin et al. [23]. Copyright 2018, Springer Nature.

5.1.3 Vapor-Phase Deposition Techniques

The small lateral size and irregular shape of 2D materials prepared by mechanical exfoliation or liquid-phase exfoliation limit the applications of 2D semiconductors in large-scale integrated optoelectronic devices. As an alternative, vapor-phase deposition techniques such as CVD and physical vapor deposition are considered as the effective methods to prepare large-scale and uniform thickness 2D semiconductors. CVD is a technique that includes three main steps: two or more kinds of raw materials were introduced into a reaction chamber, using various energy sources such as heating, plasma excitation, or optical radiation to chemically react with each other, forming materials and depositing them on a particular substrate. Until now, many kinds of 2D materials such as TMDCs, graphene, InSe have been successfully prepared by CVD on several substrates. Materials scientists, chemists, and physicists have summarized several major factors affecting the quality of thin 2D crystals prepared by CVD. (i) The supply of reactive deposits: there is no doubt, for any deposition system, that supply of the reaction mixture is one of the most important factors that determines the quality of the samples. In the process of CVD, the optimal reactant partial pressure and its relative proportion should be fully selected. (ii) Deposition temperature: it is one of the most important parameters, which directly affects the free energy of the reaction system, determines the extent and direction of the reaction, and also has a direct impact on the microstructure and chemical composition of the sample. Because of the different growth mechanisms, the degree of influence on the quality of samples is also different. In the same reaction system at different temperatures, the prepared samples may be single crystal, polycrystalline, amorphous, or even no reaction at all. (iii) Substrates: usually, the thin 2D crystal is prepared by CVD on a solid substrate and the substrate has a critical factor on the quality of the sample. The formation energy of the sample and the coefficient of expansion of the substrate will directly affect the final growth result. (iv) The pressure and gas flow of the system: in the closed tubular system, they directly affect the transport rate and have significant impact on the sample quality. For the open tubular system, the deposition is generally carried out under normal pressure, and the total force of the system is rarely considered, but in a few cases, the deposition is carried out under pressurized or reduced pressure conditions. For the vacuum system, the deposited samples usually have good uniformity and high quality. (v) The system equipment and the purity of source materials: the degree of sealing of the system, the type of chamber, and the structure of the reaction tube have an influence on the quality of the prepared samples. A large number of studies have shown that the quality of the device often results from the sample problems, and the sample quality is often directly related to the purity of the source materials. Take MoS_2 as an example, until now, many kinds of CVD methods have been developed for the synthesis of atomically thin MoS_2 on various substrates [24–26]. Independent molybdenum and sulfur precursors such as MoO_3/S, Mo/S, and Mo/H_2S were used to synthesize few-layer MoS_2. The single precursor ammonium thiomolybdate $((NH_4)_2MoS_4)$ was also used to synthesize MoS_2. The quality of MoS_2 crystals grown by these methods is quite different.

A scalable CVD method for the growth of large-area MoS_2 atomic layers on various substrates using Mo and S as the precursors is reported by Zhan group [25]. The detailed growth process is as follows: Firstly, uniform Mo films were deposited on the specific substrate using an e-beam evaporator. This substrate was placed on a ceramic boat, which was placed in the center of a tube furnace. The sulfur powders were placed in the air inlet at the end of the quartz tube, as shown in Figure 5.6a. The two sides of the quartz tube were closed. Then, air and water in the quartz tube were expelled using high-purity Ar or N_2 for about 15 minutes. Secondly, the temperature of the furnace gradually increased from 30 to 500 °C in 30 minutes, and then, the furnace was heated to 750 °C in 90 minutes and kept at 750 °C for 10 minutes. Finally, the furnace was cooled down from 750 °C to room temperature in 120 minutes. Figure 5.6b,c shows the results of the prepared MoS_2 films, which shows that the size and thickness of the MoS_2 films depend on the size of the substrate and the thickness of the pre-deposited Mo films. The results suggest that the growth process is scalable and controllable. However, the results of carrier electronic transport test show that the prepared film has a poor structural homogeneity and many defects exist in it. The obtained electron mobilities of the prepared films were ranging from 0.004

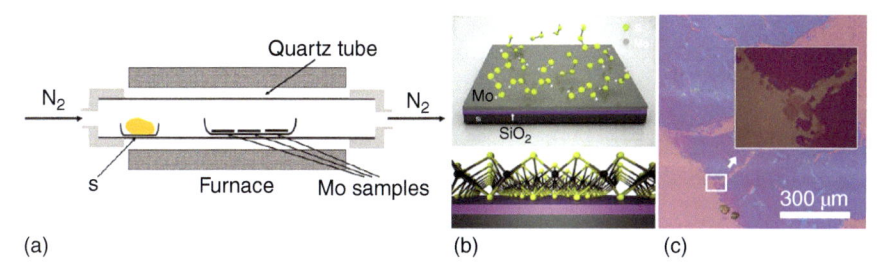

Figure 5.6 (a, b) The schematic illustration of the synthesis process. (c) Optical images of the prepared MoS$_2$ crystals. Source: Reprinted with permission from Zhan et al. [25]. Copyright 2012, John Wiley & Sons.

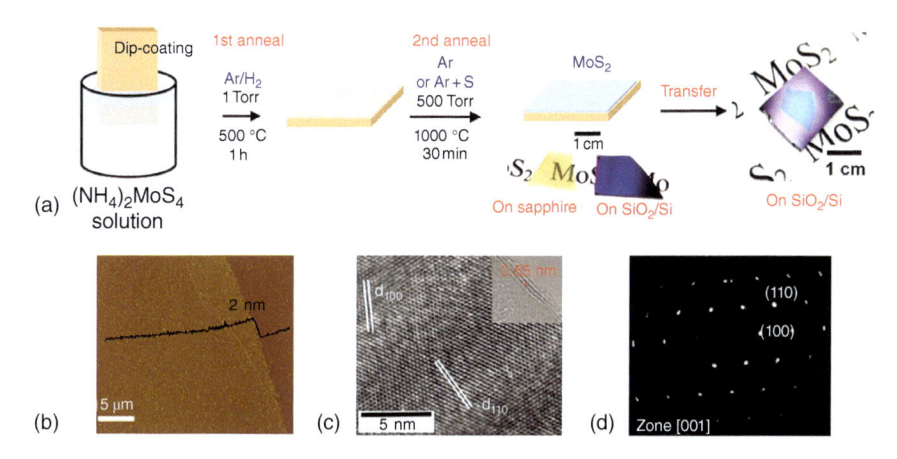

Figure 5.7 (a) Schematic illustration of the two-step thermolysis process for the synthesis of MoS$_2$ thin films. (b) AFM image of the prepared typical MoS$_2$ crystals. (c) HRTEM image of the MoS$_2$ crystal. (d) The SAED pattern of the MoS$_2$ sample. Source: Reprinted with permission from Liu et al. [26]. Copyright 2012, American Chemical Society.

to 0.04 cm^2/(V s) at room temperature, and these values are much lower than the reported values of the crystals obtained by mechanical exfoliation.

A two-step thermolysis process to synthesize large-scale MoS$_2$ thin layers has been developed by Liu et al. [26]. Figure 5.7 shows the detailed process of the growth of MoS$_2$ film. In a typical process, the clean substrates were immersed into the (NH$_4$)$_2$MoS$_4$ DMF solution; and after a few minutes, the substrate was removed from the solution to form a thin (NH$_4$)$_2$MoS$_4$ film. A heat-treat process was needed at 120 °C for 30 minutes for the (NH$_4$)$_2$MoS$_4$ film. Then, the (NH$_4$)$_2$MoS$_4$ film undergoes an annealing process in the quartz tube at 500 °C for 60 minutes. To remove the residual solvent, NH$_3$ molecules, and other by-products dissociated from the precursors, the Ar/H$_2$ atmosphere (flow rate 4 : 1) is maintained in the whole annealing process. After that, the film undergoes the second annealing process performed in Ar (or Ar + S) environment at 1000 °C. The prepared MoS$_2$ films have comparable electron mobilities with the mechanically exfoliated MoS$_2$ up to 6 cm^2/(V s). However, the process of this method is complicated, and the prepared films have many grain boundaries and defects, thus limiting its large-scale promotion.

Recent studies have found that direct sulfurization of molybdenum trioxide (MoO_3) appears to be a reliable method for preparing thin MoS_2 films. Najmaei et al. developed an improved vapor-phase MoO_3 sulfurization CVD procedure for the synthesis of large-scale MoS_2 atomic layers with high quality [24]. Firstly, highly crystalline MoO_3 nanoribbons were grown on the silicon substrates by hydrothermal method. After that, this silicon substrate covered with MoO_3 nanoribbons and several clean substrates were put into the furnace, and 0.8–1.2 g sublimated sulfur was placed in the upstream low-temperature zone of the furnace. Then, the furnace was heated from room temperature to 550 °C in 30 minutes; meanwhile, the sublimated sulfur was slowly evaporated, and then, the furnace was heated to 850 °C for 60 minutes and maintained at this temperature for 10–15 minutes. Finally, the furnace naturally cooled down to room temperature. The scanning electron microscopy (SEM) results show that the prepared large-scale MoS_2 atomic layers have the uniform and smooth surface (Figure 5.8). They also show excellent electronic properties, such as the

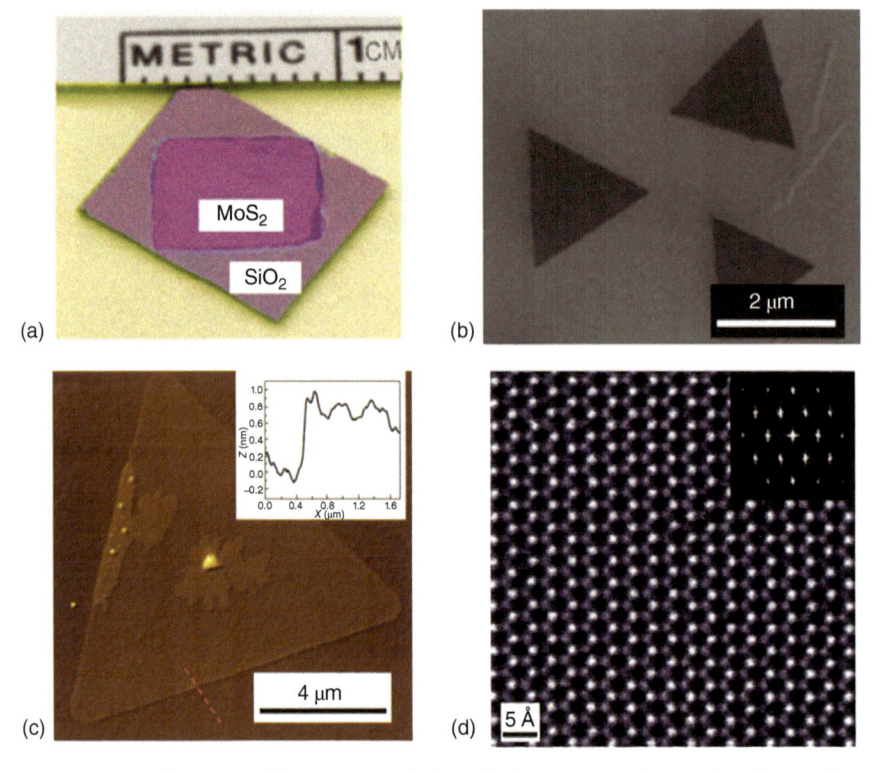

Figure 5.8 (a) The prepared large-area MoS_2 films. (b) SEM image of a typical MoS_2 sample. (c) AFM image from a MoS_2 triangular flake. (d) STEM-annular dark field (ADF) image of the triangular MoS_2 flake. Source: Reprinted with permission from Najmaei et al. [24]. Copyright 2013, Springer Nature.

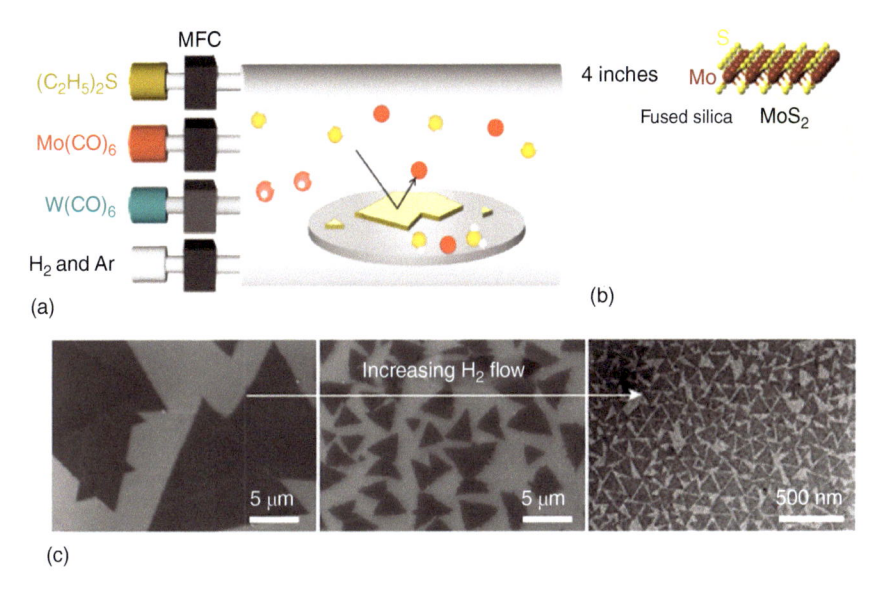

Figure 5.9 (a) Schematic illustration of the MOCVD growth setup. (b) Photographs of the prepared monolayer MoS_2 film on 4-in. fused silica substrates. (c) Grain size variation of MoS_2 crystals depending on the hydrogen flow rate; from left to right, 5, 20, and 200 standard cm^3/min (sccm). Source: Reprinted with permission from Kang et al. [27]. Copyright 2015, Springer Nature.

measured on/off ratios for the MoS_2 samples can reach to the maximum value of 6×10^6 and the carrier mobilities have an average of 3.5–5.1 $cm^2/(V\,s)$.

Kibum Kang et al. reported the preparation of high-mobility 4-in. wafer-scale films of monolayer MoS_2 and WS_2 with excellent spatial homogeneity and electrical performance, grown on SiO_2 substrates [27]. As is shown in Figure 5.9, they are grown with a metal–organic chemical vapor deposition (MOCVD) technique. $Mo(CO)_6$, $W(CO)_6$, $(C_2H_5)2S$, and H_2 were used as the gas-phase precursors, and all of these gas-phase precursors were diluted in argon as a carrier gas. They find that the layer-by-layer growth mode is the main growth mechanism of large-scale MoS_2 film with controllable uniform layer. Another important finding is that the concentrations of H_2, $(C_2H_5)2S$, as well as residual water directly affect the average grain size and the intergrain connection of MoS_2 film. Meanwhile, the films show excellent electrical properties, such as the spatial uniformity over a large scale and excellent transport properties similar to the exfoliated samples. This versatile MOCVD process provides a new avenue for the growth of high-quality monolayer semiconductor films with excellent electrical properties, which enables the realization of atomically thin integrated circuitry.

Very recently, Zheng Liu group prepared a wide variety of two-dimensional monolayer TMDCs by the molten-salt-assisted CVD [28]. The results show that the competition between the mass flux of the metal precursors and the reaction rate of the domains directly affects the morphology of the samples. The mass flux affects the amount of metal precursors, which plays a major role in the formation

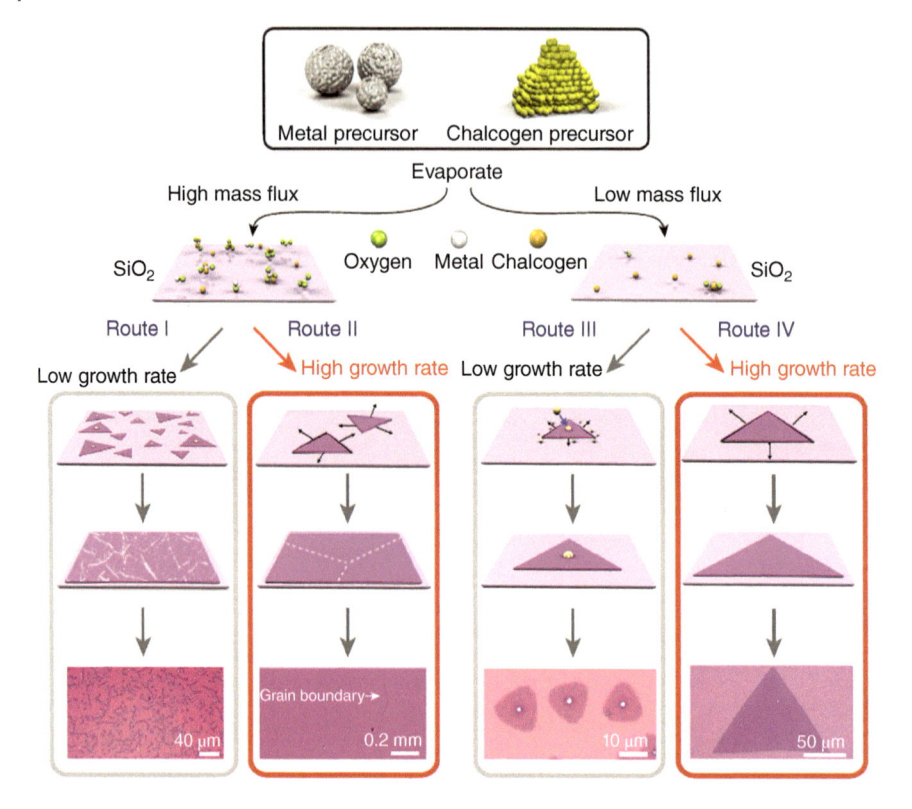

Figure 5.10 Flow chart of the general growth process for the production of TMCs by the chemical vapor deposition method. Source: Reprinted with permission from Zhou et al. [28]. Copyright 2018, Springer Nature.

of nucleus and the growth of domains, and the growth rate determines the grain size of the final samples. As is shown in Figure 5.10, the large mass flux and low growth rate result in the formation of a monolayer sample with small grains (route I), and high growth rate leads to the large-scale monolayer samples with the grain size up to millimeters (route II). However, the small mass flux and low growth rate tend to form small flake samples. At the center of the samples, the tiny nuclei are often observed, which suggests that in the whole growth, the extra adatoms or atom clusters always attach to the existing nucleus or to the edge (route III). In addition, the high reaction rate is very helpful for producing the monolayer of individual large 2D single crystals (route IV). Using this method, they synthesized a library of good-quality 2D transitional metal chalcogenides (TMCs).

5.2 Characterizations of 2D Semiconductors

5.2.1 Surface Morphology (SEM, OM, and TEM)

SEM produces images of samples by scanning the surface with a focused beam of electrons, which is an effective tool to characterize the microscopic morphology of 2D materials. In the process where the electrons interact with atoms, various signals that contain the surface topography and composition information about the samples will be produced. These signals are then detected and the magnified image is obtained by point-by-point scanning. Among these signals are mainly the secondary electrons of the sample, which are emitted by atoms excited by the electron beam. In addition, the secondary electrons are detected using the Everhart–Thornley detector. The number of detected secondary electrons can be used to analyze the morphological information of the samples. SEM is widely used in the study of characterizing the grain size, grain morphology, and grain orientation of 2D materials because of the advantages of high magnification, large depth of field, high resolution, etc. Shanshan Wang et al. prepared monolayer MoS_2 two-dimensional crystals at elevated temperatures on silicon substrates with a 285 nm oxide layer by atmospheric pressure CVD [29]. They also studied the effects of growth temperature, gas flow, and Mo : S ratio on the shape evolution of monolayer MoS_2 crystals. As is shown in Figure 5.11, along the Ar gas flow direction at 750 °C on the Si chip, the MoS_2 shape transformation phenomenon widely exists. This is mainly because the temperature influences the evaporation amount of MoO_3, which leads to the presence of the different MoO_3 concentration gradients in the gas phase. Figure 5.11 shows the MoS_2 shape transformation along the Ar gas flow direction at 750 °C on the third Si chip.

As a means of characterization of macroscopic materials, optical microscope (OM) cannot give crystal structure information of two-dimensional materials, but in special cases, it provides a fast and convenient method for layer number determination and morphology analysis of two-dimensional materials. In fact, OM enables rapid detection of number of layers for two-dimensional materials before using more precise methods for thickness detection such as atomic force microscopy and transmission electron microscopy. It is generally believed that the visibility of 2D material under the OM results from the change in interference color of 2D material was compared to the blank substrate. When we use

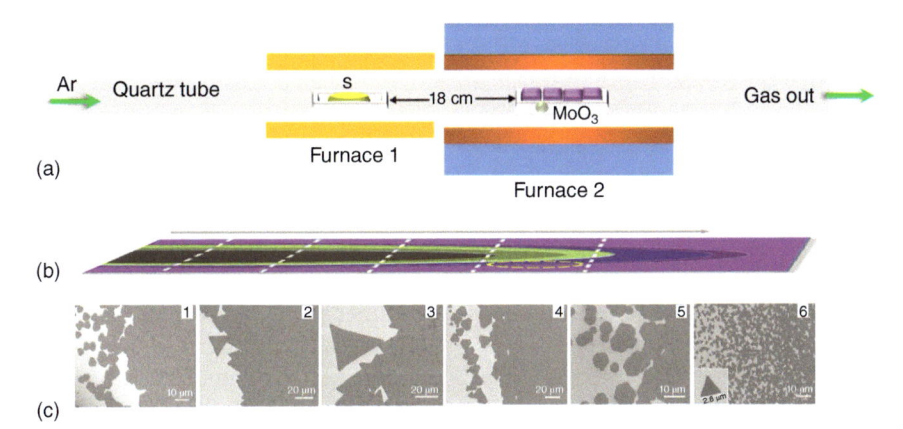

Figure 5.11 (a) Schematic illustration of the MoS_2 CVD system. (b, c) Schematic illustration showing the spatial sectioning of the growth substrate into sections 1–6 and the corresponding SEM images from the edge of those sections. Source: Reprinted with permission from Wang et al. [29]. Copyright 2014, American Chemical Society.

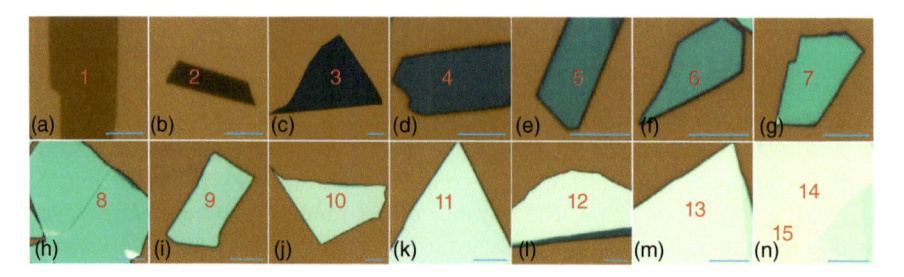

Figure 5.12 Color OMs of the MoS$_2$ nanosheets from 1L to 15L on 90 nm SiO$_2$/Si substrate. Scale bar is 5 μm. The layer numbers of corresponding MoS$_2$ nanosheets are marked in the corresponding images. Source: Reprinted with permission from Li et al. [30]. Copyright 2013, American Chemical Society.

the OM to observe the monolayer or few-layer 2D materials, we find that different layers of 2D materials have different colors. The supporting substrate of 2D materials and the OM components also have a crucial influence on the observation of monolayer or few-layer 2D materials. As early as 2012, Zhang Hua's group from Nanyang Technological University proposed a simple, fast, and reliable method for determining the number of layers of two-dimensional materials [30]. They selected SiO$_2$/Si (the thicknesses of SiO$_2$ are 90 and 300 nm, respectively) as the substrates and then made a standard table based on the contrast difference between the two-dimensional material of different thicknesses and the substrates. Therefore, we can quickly obtain the thickness of the two-dimensional material by comparing the table. Figure 5.12 shows the OM contrast change of the MoS$_2$ flakes with different thicknesses on the SiO$_2$/Si (the thickness of SiO$_2$ is 90 nm) substrates. They show that 1L–15L MoS$_2$ flakes on 300 nm SiO$_2$/Si can be reliably identified by measuring the contrast difference (defined as C_D) value in combination with the contrast difference values in the gray-scale image (from the R, G, or B channel, C_{DR}, C_{DG}, and C_{DB}). The MoS$_2$ flakes with layers from 7L to 15L can be rapidly and reliably identified by comparing the C_D values. However, it is difficult to distinguish the samples with 1L–6L. They further found that the samples with the layer from 1L to 6L can be easily judged by C_{DR}, C_{DG}, and C_{DB}. Therefore, the MoS$_2$ flakes with 1L–15L on SiO$_2$/Si substrates can be readily identified using the C_D values in combination with the C_{DB}, C_{DR}, and C_{DG} values.

5.2.2 Thickness (Raman, AFM, and HRTEM)

Accurately determining the thickness of two-dimensional layered materials has an important impact on the physical and chemical properties of two-dimensional materials and their applications. AFM is an analytical instrument that directly determines the surface information of a sample. According to the type of force between the tip and the sample, the working mode of the AFM is usually divided into three types: contact mode, noncontact mode, and tapping mode. For the contact mode, the distance between the tip and the sample is relatively <0.5 nm, and the tip moves on the surface of the sample. At this time, the force between the tip atom and the sample atom is van der Waals repulsive force interaction. This contact mode can achieve a higher resolution up to 0.1 nm in the lateral direction, but it can damage the surface of the sample. For the noncontact mode, the tip is always scanned at a position far from the surface of the sample. The interaction of the attraction between the tip and the sample atom is used to judge the surface information of the sample. This mode does not cause damage to the sample, but it has a lower resolution. For the tapping mode, the surface topography image information is recorded by the amplitude of the up and down vibration of the probe. Specifically, it is to drive the probe arm to vibrate near its resonance frequency by the piezoelectric piece and adjust the distance between the sample and the tip to control the amplitude of the arm. The tap mode not only enables high-resolution scanning but also does not damage the sample. In the study of two-dimensional layered materials, the tapping mode AFM is often used to directly judge the thickness information of the sample.

In addition, Raman spectroscopy is an effective method to judge the number of layers of 2D materials, and this in-direct method of determining the number of layers has been widely used. Raman spectroscopy is a spectroscopy technique based on inelastic scattering of monochromatic light. The analysis of scattered light with different incident light frequencies can obtain vibration and rotation information of the sample and then analyze the physical information. Studies have shown that the peak position and intensity of the Raman peak for two-dimensional materials will vary greatly with the thickness of the sample. For example, the recent reported 2D semiconductor-black arsenic shows obvious thickness-dependent optical properties [31]. Figure 5.13a shows the Raman spectra of black arsenic crystals with monolayer, few-layer, and bulk prepared by the mechanically exfoliated method. It can be seen that two strong peaks at 223.6 and 251 cm^{-1} are observed in bulk b-As, and these peaks correspond to the B_{2g} and A_{2g} vibration modes, respectively. It can also be found that blue shifts of A_g and E_g modes were observed when the thickness of b-As decreases from bulk to monolayer. In addition, the intensity of these two modes gradually weakened as the decrease of sample thickness.

5.2.3 Phase Structure (HRTEM and STEM)

X-ray diffraction (XRD) is the simplest and quickest test technique for judging the phase structure of materials; however, the XRD technology is no longer suitable for few layers (especially for the monolayer) of two-dimensional layered materials. For the few-layer two-dimensional layered materials, HRTEM is a widely recognized technology to characterize the phase structure. According to the results of HRTEM, the surface atom stack information of the few-layer two-dimensional layered materials can be obtained to judge the phase structure of them. Scanning transmission electron microscopy (STEM), as a higher resolution technology, can achieve a more accurate characterization of the phase structure. Figure 5.13b,d shows the typical HRTEM results of 2H and 1T MoS_2, and Figure 5.13c exhibits the SAED image of 1T MoS_2 [32, 33]. They show a clear difference in atomic stacking. Therefore, HRTEM or STEM is the important technique for judging the phase structure of materials.

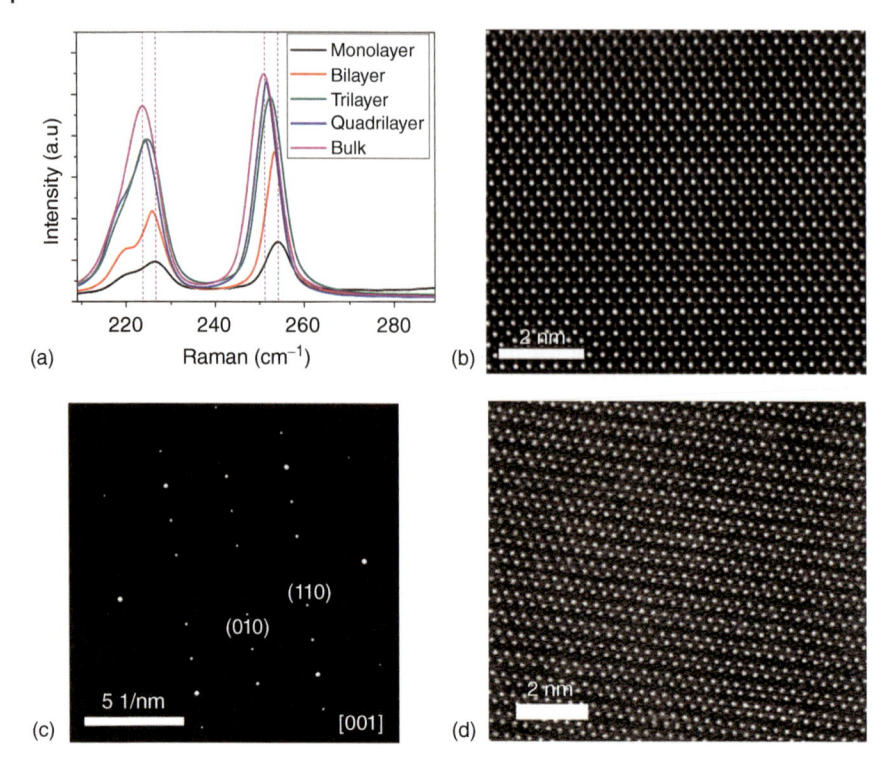

Figure 5.13 (a) Micro-Raman spectra of monolayer and multilayer (1L to bulk) b-As flakes. Source: Reprinted with permission from Zhong et al. [31]. Copyright 2018, John Wiley & Sons. (b) STEM-ADF image of a typical monolayer MoS_2. Scale bar is 2 nm. Source: Reprinted with permission from Zhong et al. [32]. Copyright 2019, Springer Nature. (c, d) SAED and highangle annular dark field (HAADF) image of 1T'-MoS_2 monolayer. Source: Reprinted with permission from Peng et al. [33]. Copyright 2019, John Wiley & Sons.

5.2.4 Band Structure (Optical Absorption and Photoluminescence, ARPES)

The electronic band structure of the two-dimensional layered material has a strong thickness. The optical absorption and photoluminescence can indirectly determine the band gap and the type of the band structure. These two characterization methods are simple and the test results can be fast obtained. Angle-resolved photoemission spectroscopy (ARPES) is an efficient technique to study the electronic band structure of materials using photoelectric effects. However, the equipment used in this method is expensive, and it also has a long test period; currently, fewer laboratories can implement the test. The electronic band structures of a large number of two-dimensional layered materials have been tested by ARPES [34].

5.2.5 Chemical Composition and Chemical States (XPS and EDS)

X-ray photoelectron spectroscopy (**XPS**) is performed when a material is irradiated with X-rays, and the electrons of atoms or molecules in the sample are excited to escape from the surface of the material being analyzed and then to measure the number of electrons and the kinetic energy. Typically, the XPS spectrum is a plot of the number of electrons detected versus the binding energy of the electrons detected. Each element produces a set of characteristic XPS peaks at the characteristic binding energy values that directly identify each element present in or on the surface of the material being analyzed. These characteristic spectral peaks correspond to the electronic configuration of electrons within the atom. The number of electrons detected in each characteristic peak is directly related to the amount of elements in the XPS sample volume. XPS can measure the elemental composition, chemical state, electronic state, and the elemental empirical formula that exist within a material. Therefore, it is a useful surface-sensitive quantitative spectroscopic measurement technique [35].

Energy-dispersive X-ray spectroscopy (EDS or EDX) is a useful analytical technique to analyze the elements or chemical characterization of a sample. In addition, it is an important technique to quickly analyze the types and contents of the elements of the material being analyzed. It relies on the fundamental principle that each element has a unique atomic structure allowing for a unique set of peaks in its electromagnetic emission spectrum. Usually, the high-energy beam of charged particles such as a beam of X-rays or electrons is focused into the sample to excite the elements to produce the characteristic X-rays. Then, the number and energy of the X-rays emitted from a sample were measured by an

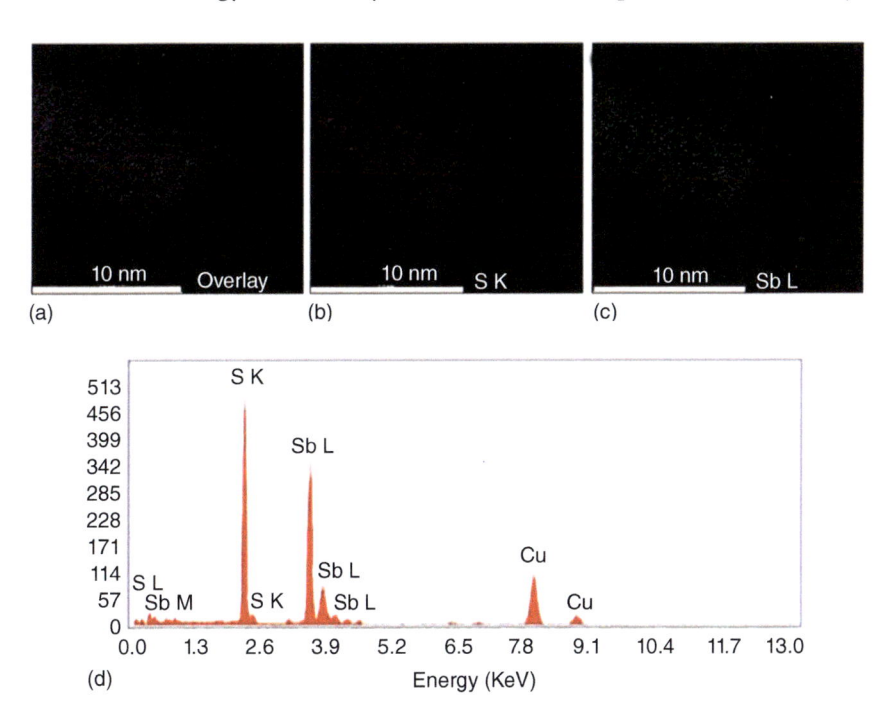

Figure 5.14 (a–c) HAADF STEM-EDX mappings of an individual nanowire reveal uniform elemental distributions. (d) The representative EDX spectrum of a single Sb_2S_3 nanowire. The Cu signal comes from the Cu grid. Source: Reprinted with permission from Zhong et al. [36]. Copyright 2017, Royal Society of Chemistry.

energy-dispersive spectrometer. The energies of the X-rays are the characteristic of the atomic structure of the emitting element and the difference in energy between the shells, so, for a specific element, it has a defined X-ray characteristic energy. Usually, EDS or EDX is used as an accessory for transmission electron microscope (TEM), scanning electron microscopes (SEMs), and scanning transmission electron microscopes (STEMs); it is capable of providing samples in high spatial resolution for chemical determination and content determination in microzone. In the practical application, three sampling methods of point, line, and surface are often used to provide three kinds of results: characteristic peak spectrum, line scan, and element surface analysis. The EDX elemental mappings of the typical Sb_2S_3 NW are shown in Figure 5.14 [36]. It can be seen that the Sb and S atoms are homogeneously distributed across the whole crystal. Figure 5.14 shows the collected EDX spectrum from a point of crystal. As shown in Figure 5.14, the crystal is composed of S and Sb and the stoichiometric ratio of Sb and S is about 2 : 3.

5.3 Electrochemical Properties of 2D Semiconductors

Electrochemistry, as a measurable and quantitative phenomenon, as well as identifiable chemical changes, mainly studies the relationship between electricity [37]. It deals with the interaction between electrical energy and chemical change. An electrochemical reaction is a chemical reaction caused by an externally supplied current or a current generated by a spontaneous chemical reaction. In the process of these reactions, they involve electric charges moving between electrolyte and electrodes. Electrocatalysts are an important reactive species required during electrochemical reactions. The electrocatalyst is used to increase the rate of the chemical reaction and is not itself consumed in the chemical reaction. A wide variety of electrocatalysts have been developed, but high-efficiency catalysts are yet to be further developed. Finding effective ways to regulate interface charge transfer is a productive way to accelerate heterogeneous catalysis. It is believed that the thermodynamic driving force and kinetic approach of adjusting the heterogeneous charge transfer process at the semiconductor surface is an effective strategy to improve the catalytic efficiency of electrochemical reactions. The chemical or ion modification, defect control, surface passivation, etc., on the surface of the semiconductor are considered the efficient methods to improve the catalytic efficiency of electrochemical reactions. In addition, the catalyst material also has important impact on improving the electrochemical reaction. Because of the unique crystal structure of the two-dimensional layered materials, they have the important application prospect in electrochemical catalysis. $2H-MoS_2$ has the band gap of 1.2–1.9 eV (from bulk to monolayer) and strongly absorbs light in the visible spectrum making possible direct photoelectrocatalysis [38]. Studies also find that the 1T-phase of MoS_2 has excellent properties in the hydrogen evolution reaction (HER) [39–42].

References

1 Cui, X., Lee, G.-H., Kim, Y.D. et al. (2015). *Nat. Nanotechnol.* 10: 534.
2 Tongay, S., Sahin, H., Ko, C. et al. (2014). *Nat. Commun.* 5: 3252.
3 Tongay, S., Zhou, J., Ataca, C. et al. (2012). *Nano Lett.* 12: 5576.
4 Yin, Z., Li, H., Li, H. et al. (2011). *ACS Nano* 6: 74.
5 Najmaei, S., Liu, Z., Ajayan, P., and Lou, J. (2012). *Appl. Phys. Lett.* 100: 013106.
6 Plechinger, G., Nagler, P., Kraus, J. et al. (2015). *Phys. Status Solidi Rapid Res. Lett.* 9: 457.
7 Koperski, M., Nogajewski, K., Arora, A. et al. (2015). *Nat. Nanotechnol.* 10: 503.
8 Wang, Y., Liu, E., Liu, H. et al. (2016). *Nat. Commun.* 7: 13142.
9 Guo, S., Yang, D., Zhang, S. et al. (2019). *Adv. Funct. Mater.*: 1900138.
10 Keum, D.H., Cho, S., Kim, J.H. et al. (2015). *Nat. Phys.* 11: 482.
11 Huang, Y., Sutter, E., Shi, N.N. et al. (2015). *ACS Nano* 9: 10612.
12 Desai, S.B., Madhvapathy, S.R., Amani, M. et al. (2016). *Adv. Mater.* 28: 4053.
13 Carey, B.J., Daeneke, T., Nguyen, E.P. et al. (2015). *Chem. Commun.* 51: 3770.
14 Smith, R.J., King, P.J., Lotya, M. et al. (2011). *Adv. Mater.* 23: 3944.
15 Hernandez, Y., Nicolosi, V., Lotya, M. et al. (2008). *Nat. Nanotechnol.* 3: 563.
16 Jawaid, A., Nepal, D., Park, K. et al. (2015). *Chem. Mater.* 28: 337.
17 Yasaei, P., Kumar, B., Foroozan, T. et al. (2015). *Adv. Mater.* 27: 1887.
18 Li, Z., Qiao, H., Guo, Z. et al. (2018). *Adv. Funct. Mater.* 28: 1705237.
19 Yang, Z., Liang, H., Wang, X. et al. (2015). *ACS Nano* 10: 755.
20 Knirsch, K.C., Berner, N.C., Nerl, H.C. et al. (2015). *ACS Nano* 9: 6018.
21 Zeng, Z., Sun, T., Zhu, J. et al. (2012). *Angew. Chem. Int. Ed.* 51: 9052.
22 Song, H., Li, T., Zhang, J. et al. (2017). *Adv. Mater.* 29: 1700441.
23 Lin, Z., Liu, Y., Halim, U. et al. (2018). *Nature* 562: 254.
24 Najmaei, S., Liu, Z., Zhou, W. et al. (2013). *Nat. Mater.* 12: 754.
25 Zhan, Y., Liu, Z., Najmaei, S. et al. (2012). *Small* 8: 966.
26 Liu, K.-K., Zhang, W., Lee, Y.-H. et al. (2012). *Nano Lett.* 12: 1538.
27 Kang, K., Xie, S., Huang, L. et al. (2015). *Nature* 520: 656.
28 Zhou, J., Lin, J., Huang, X. et al. (2018). *Nature* 556: 355.
29 Wang, S., Rong, Y., Fan, Y. et al. (2014). *Chem. Mater.* 26: 6371.
30 Li, H., Wu, J., Huang, X. et al. (2013). *ACS Nano* 7: 10344.
31 Zhong, M., Xia, Q., Pan, L. et al. (2018). *Adv. Funct. Mater.* 28: 1802581.
32 Zhong, M., Shen, C., Huang, L. et al. (2019). *npj 2D Mater. Appl.* 3: 1.
33 Peng, J., Liu, Y., Luo, X. et al. (2019). *Adv. Mater.*: 1900568.
34 Mahatha, S., Patel, K., and Menon, K.S. (2012). *J. Phys. Condens. Matter* 24: 475504.
35 Liu, J., Zhong, M., Liu, X. et al. (2018). *Nanotechnology* 29: 474002.
36 Zhong, M., Wang, X., Liu, S. et al. (2017). *Nanoscale* 9: 12364.
37 Velický, M. and Toth, P.S. (2017). *Appl. Mater. Today* 8: 68.
38 Wang, Y., Ou, J.Z., Balendhran, S. et al. (2013). *ACS Nano* 7: 10083.

39 Wang, D., Xiao, Y., Luo, X. et al. (2017). *ACS Sustain. Chem. Eng.* 5: 2509.
40 Lukowski, M.A., Daniel, A.S., Meng, F. et al. (2013). *J. Am. Chem. Soc.* 135: 10274.
41 Tang, Q. and Jiang, D.-e. (2016). *ACS Catal.* 6: 4953.
42 Chang, K., Hai, X., Pang, H. et al. (2016). *Adv. Mater.* 28: 10033.

6

Properties of 2D Alloying and Doping

6.1 Introduction

Two-dimensional materials have increasingly important research significance because of their excellent physical properties, as well as extraordinary practical applications in optical and optoelectronic fields. The unique lattice structure makes the two-dimensional materials have better controllable properties.

In order to obtain more suitable bandgap and carrier transport materials in the fields of electronics, nanoelectronics, and photonics, as well as the need for tunable spectral response, it is very meaningful to realize the bandgap tunability of two-dimensional semiconductor materials. Alloying is a very important method to realize the adjustable bandgap of two-dimensional materials [1–3]. In other words, alloys provide a means to tailor specific spectral responses and device characteristics [4]. Moreover, the electronic structure and lattice parameters of the material can be better adjusted by alloying, which is not only valuable for materials but also the device is important; the flexible regulation of material properties makes it possible to manufacture a wider range of applications and high-performance devices.

6.2 Advantages of 2D Alloys

The method of alloying is one of the widest methods for regulating the properties of two-dimensional materials. The two-dimensional alloy is a novel two-dimensional material formed by substitutional doping. Substitutional doping is to replace one of the atoms in the crystal with another atom. From the perspective of the semiconductor energy band, the alloy changes the structure of conduction band and valence band by regulating the proportion of elements [2]. Controllable two-dimensional semiconductor materials bring more possibilities to the development and application of high-performance devices. Figure 6.1 is a schematic diagram of heterojunction and alloying. Unlike heterojunctions, which rely on two or more layers of materials, the alloy provides more flexible regulation and more possibilities.

For different two-dimensional materials, studies show that boron carbon nitride (BCN) materials are thermodynamically unstable [6, 7], but theoretical

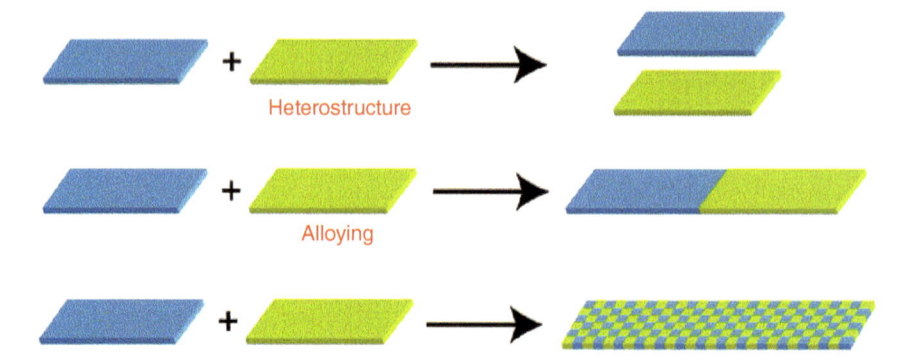

Figure 6.1 Schematic of tuning 2D TMD properties by constructing heterostructures and alloying. Source: Reproduced with permission from Wang et al. [5]. Copyright 1972, Royal Society of Chemistry.

calculation shows that the mixed ternary $MoS_2/MoSe_2/MoTe_2$ 2D compounds are thermodynamically stable at room temperature [8], which proves that transition metal dichalcogenide (TMDC)s can be used to experimentally synthesize alloys. For TMDCs, most TMDCs have a similar structure and lattice constants, which give them more possibilities for fusion. Moreover, for different TMDC materials, they have very different bandgaps, and the properties vary from metallic to wide bandgap. The properties of TMDCs provide more choices to 2D semiconductor alloys.

For two-dimensional materials, especially for TMDCs, in order to be widely used in the field of optoelectronics, the wide spectral response caused by adjustable bandgap is very important and significant. It is an effective way to make semiconductor alloys with adjustable composition and to control the electronic and optical properties of materials by controlling their chemical composition and physical size [4]. By changing the composition of the material, the electronic and optical properties of the materials can be tuned to applications with some specific conditions, such as solar cells, radiation detectors, or gas sensors [7].

6.2.1 Adjustable Bandgap

The bandgap adjustability is a very important aspect for semiconductor materials and device applications. By controlling the group distribution ratio in the alloy semiconductor and the control of the number of layers of the alloy material, two-dimensional materials of alloy semiconductors with different properties can be obtained. Coupled with the construction of heterostructures of two-dimensional alloy semiconductor materials, more electronic and photonic devices with excellent performance can be obtained, which greatly expands the potential application of two-dimensional materials in nanoelectronics and nanophotonics [9].

Different two-dimensional materials have different bandgaps. These materials can be used to synthesize alloys with different ratios of different bandgaps [9].

This method is of great significance for the application of two-dimensional materials in optical and electrical fields. Many people have studied the two-dimensional alloy theoretically and experimentally. In theory, they mainly research the electronic structure [8, 10, 11], thermodynamic stability [8, 12], and kinetic stability of the alloy. The experiments mainly studied the alloy synthesis, various properties of the alloy, performance of the device, etc. For many two-dimensional materials, TMDCs are the most studied in terms of two-dimensional alloys.

The TMDCs themselves have great advantages, and the alloys with the two-dimensional family bring more possibilities. By forming alloys of two or more TMDCs, the TMDCs alloy has the characteristics that the band structure (bandgap) is continuously adjustable. Most of the early studies were MoS_2- and WS_2-doped metal elements (W and Mo) to obtain $Mo_xW_{1-x}S_2$ and $Mo_{1-x}W_xS_2$ alloys, doped with chalcogen (Se) to obtain $MoS_{2(1-x)}Se_{2x}$, $WS_{2(1-x)}Se_{2x}$, and similar selenium(Se) doping. Subsequently, more and more excellent 2D materials were discovered and prepared, and the exploration of alloys was more diversified, making more 2D semiconductor alloys researched and prepared.

The following is a description of specific alloys.

For example, in the synthesis of $WS_{2x}Se_{2-2x}$ alloys, an alloy of different ratios of S atoms and Se atoms is obtained by controlling the group distribution ratio in the material. The limit of the ratio of S atom to Se atom is pure WS_2 and WSe_2. The difference in bandgap between the two materials also indicates the tunability of bandgap in $WS_{2x}Se_{2-2x}$ alloy material, with the difference of S/Se ratio ($0 \leq x \leq 1$) basically presenting a linear relationship (Figure 6.2a) [4].

In addition, from the perspective of formation energy, $MoSe_{2-2x}S_{2x}$ is thermodynamically stable using theoretical calculations [10]. The calculation results also show that the bandgap of $Mo_{1-x}W_xS_2$ varies between 1.87 and 2.0 eV, and the bandgap of $MoSe_{2(1-x)}S_{2x}$ varies from 1.62 to 1.86 eV (Figure 6.2b). It can be seen that this change still meets the variation of the two limits, and the variation of $MoSe_{2(1-x)}S_{2x}$ is close to linear. The first-principles calculation has great guiding significance for finding more viable alloys with excellent properties. Theoretical calculations and experimental studies have jointly promoted the rapid development of 2D semiconductor alloys.

In order to obtain more tunable bandgap alloys to suit the needs of different bandgaps, Fangfang Cui et al. consider the bandgap of single-layer ReS_2 and $ReSe_2$ (1.6 eV, 1.3 eV, respectively) and properties and synthesize 1T'-phase $ReS_{2x}Se_{2(1-x)}$ monolayer film by chemical vapor deposition (CVD) [15]. The bandgap of $ReS_{2x}Se_{2(1-x)}$ varies with S composition from 1.32 to 1.62 eV (Figure 6.2c), which provides a choice for two-dimensional materials in the near-infrared range (NIR). Later, Wen Wen et al. considered that rhenium dichalcogenides (ReX_2; X = S, Se) had lower lattice symmetry and produced inherent in-plane anisotropy, which made ReX_2 have new application potential. The $ReS_{2(1-x)}Se_{2x}$ alloy was synthesized by chemical vapor transport (CVT) [16]. The tunable bandgap range is from 1.31 to 1.62 eV (consistent with the work of Fangfang Cui and coworkers [15]). Furthermore, Yucheng Huang et al. calculated the bandgap adjustment (nonlinear relationship) of $SnSe_{2(1-x)}S_{2x}$ alloy from 1.14 to 2.08 eV by first-principles calculation [17]. After that, Yan Wang et al.

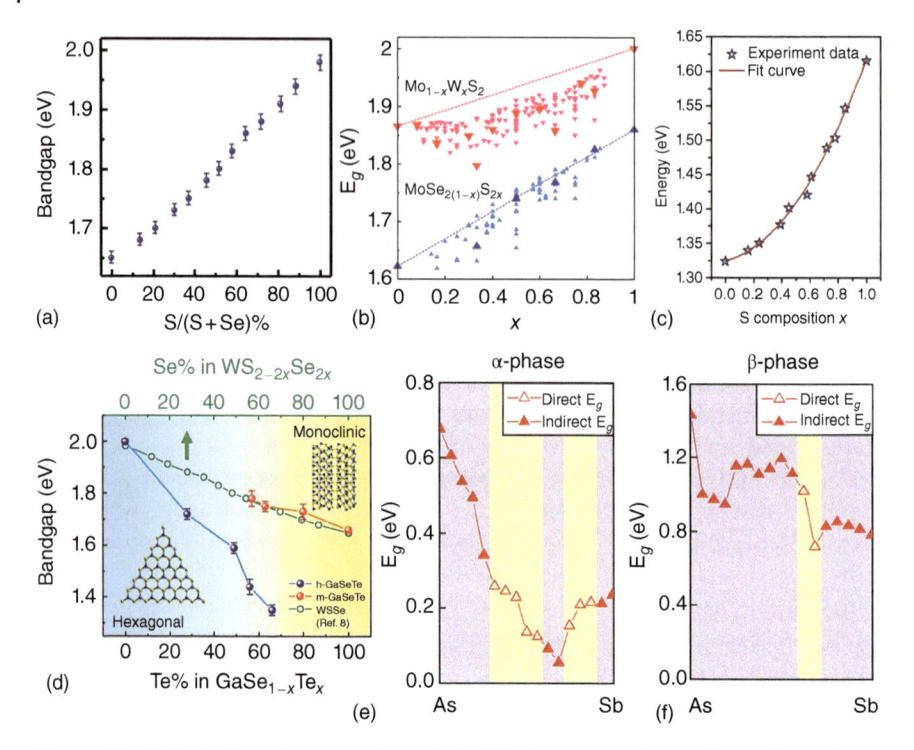

Figure 6.2 (a) Optical bandgap vs. sulfur ratio in $WS_{2x}Se_{2-2x}$ nanosheets. Source: Reproduced with permission from Duan et al. [4]. Copyright 2016, American Chemical Society. (b) The local-density approximation (LDA) bandgap in $Mo_{1-x}W_xS_2$ and $MoSe_{2(1-x)}S_{2x}$ alloys as a function of concentration x. The calculation method is calculated by density functional theory. Source: Reproduced with permission from Kutana et al. [10]. Copyright 2009, Royal Society of Chemistry. (c) E_g as a function with the S component of $ReS_{2x}Se_{2(1-x)}$ alloy. Source: Reproduced with permission from Cai et al. [15]. Copyright 2017, Springer Nature. (d) Bandgap as a function of Te content in $GaSe_{1-x}Te_x$. (e) $\alpha\text{-}As_xSb_y$ and (f) $\beta\text{-}As_xSb_y$ bandgap as a function of composition. Source: Reproduced with permission from Zhao et al. [14]. Copyright 2013, Royal Society of Chemistry.

successfully prepared $SnSe_{2(1-x)}S_{2x}$ crystals with different compositions using CVT and fabricated monolayer nanosheets by mechanical exfoliation [18]. Jing Xia et al. prepared $CdS_xSe_{(1-x)}$ alloys by CVD and obtained bandgap tuning from 1.8 to 2.3 eV. This result is also consistent with their density functional theory (DFT) calculation results [19]. Denggui Wang et al. reported for the first time that $HfS_{2(1-x)}Se_{2x}$ alloy was synthesized over a large area, and the continuously adjustable composition has a bandgap tuning from 1.94 to 2.64 eV [20].

Materials of the same phase can be synthesized as alloys of the stationary phase, and two completely different materials can also be alloyed. Hui Cai et al. used the isotropic hexagonal GaSe and the anisotropic monoclinic GaTe to synthesize the $GaSe_{1-x}Te_x$ alloy, and obtained a wide bandgap tuning range of 1.35–2.01 eV (Figure 6.2d). In addition, because of the existence of different phases, it also results in more possibilities to this two-dimensional alloy [13].

Along with the enumerated alloys, there is still a lot of work to study the tunable bandgap of 2D semiconductor alloys.

Kuc et al. calculated the stability and electronic structure of the phase for the $Mo_{1-x}X_xS_{2-y}Se_y$ (X = W, Nb) monolayer alloys [11]. The structure uses MoS_2 to replace Mo atoms with W and Nb atoms and Se atoms to replace S atoms. The results show that the $Mo_{1-x}Nb_xS_2$ alloy is difficult to achieve in terms of energy and structure, while W and Se are easily mixed with MoS_2. The $Mo_{1-x}W_xS_2$ monolayer alloys achieve bandgap tuning from 1.84 to 2.01 eV. Yanfeng Chen et al. synthesized a single layer of $Mo_{1-x}W_xS_2$ alloys and obtained a continuous tuning bandgap from 1.82 eV ($x = 0.20$) to 1.99 eV ($x = 1$) [21]. Q. Feng et al. grew different compositions of $MoS_{2(1-x)}Se_{2x}$ alloys ($x = 0$–0.40 and $x = 0.41$–1.00), and the resulting bandgap tuning ranged from 1.55 to 1.86 eV [13, 22, 23]. Hyung Soon Im et al. synthesized $Sn_xGe_{1-x}S$, $Sn_xGe_{1-x}Se$, GeS_xSe_{1-x}, and SnS_xSe_{1-x} alloys (where $x = 0$–1), and the alloys of these four orthogonal phases can achieve a wide range of bandgap tuning from 0.9 to 1.6 eV [24].

In addition to the extensive research of TMDCs, the new two-dimensional materials – arsenic and terpenes – have also attracted the attention of some researchers. N. Zhao et al. selected the most stable phase α and phase β for research [14]. First, phonon spectra and molecular dynamics simulations show that they have good thermodynamic stability and kinetic stability. Theoretical calculations show that α-As_xSb_y can achieve bandgap modulation from 0.06 to 0.68 eV and β-As_xSb_y from 0.72 to 1.43 eV with bandgap modulation (Figure 6.2e,f). In addition, the transition from the indirect bandgap to the direct bandgap is achieved in alloys of certain components. By the way, the effective quality has also changed with the bandgap modulation. In addition, John Mann et al. synthesized $MoS_{2(1-x)}Se_{2x}$ monolayer alloys to achieve bandgap tuning from 1.55 to 1.87 eV [25].

6.2.2 Carrier-Type Modulation

In addition to bandgap modulation, which is a very important aspect of 2D semiconductor alloys, carrier-type modulation is another important aspect of alloying in optoelectronic applications.

Xufan Li et al. synthesized a $Mo_{1-x}W_xSe_2$ ($x = 0$–0.18) monolayer alloy and achieved carrier-type modulation [26]. With the increase of W concentration, the n-type characteristics of $MoSe_2$ are inhibited, and the $Mo_{1-x}W_xSe_2$ monolayer alloy gradually enhances its p-type characteristics (Figure 6.3a). Finally, PN junctions are fabricated by stacking p-type $Mo_{1-x}W_xSe_2$ with n-type $MoSe_2$. The test results show good current rectification characteristics and gate voltage regulation characteristics [26]. This method of controlling the conductive type of materials by alloys provides a new idea for the application of two-dimensional materials [26, 27].

Fangfang Cui et al. studied carrier modulation in $ReS_{2x}Se_{2(1-x)}$ single-layer alloys, and the carrier type has a significant dependence on the alloy composition (Figure 6.3b) [15]. The p-type semiconductor characteristics are exhibited for the Se-rich alloy ($x = 0$–0.3), and the n-type semiconductor characteristics are mainly exhibited for the S-rich alloy ($x = 0.6$–1.0). The alloy of the composition

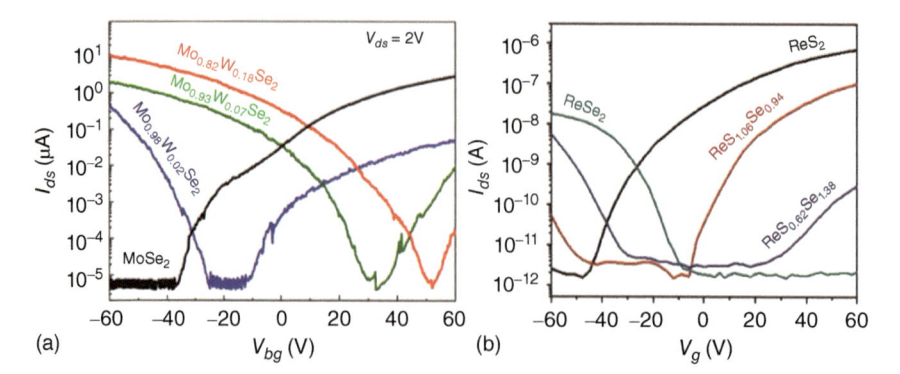

Figure 6.3 (a) Transfer curves ($V_{ds} = 2$ V) of FETs based on monolayer MoSe$_2$ (black) and Mo$_{1-x}$W$_x$Se$_2$ monolayer alloys with different W concentrations (blue, green, and red for $x = 0.02, 0.07$, and 0.18, respectively). Source: Reproduced with permission from Li et al. [26]. Copyright 2016, John Wiley & Sons. (b) Transfer curves of several representative components (with x different) of ReS$_{2x}$Se$_{2(1-x)}$ alloy transistors. Source: Reproduced with permission from Cui et al. [15]. Copyright 2017, John Wiley & Sons.

between them ($x = 0.3$–0.6) exhibits significant bipolar properties. In addition, the threshold voltage and carrier mobility are also modulated by the alloy composition.

Alloys with continuously varying compositional components can be seen with gradual changes in their n-type and p-type properties. In addition, the synthesis of alloys of fixed components and comparison with intrinsic materials is also a method of tuning the carrier type.

6.2.3 Phase Change

In addition to the excellent performance of 2D alloys in optical/electrical devices, 2D materials also demonstrate potential applications in other areas, such as phase change memory (PCM). Metal–semiconductor phase transition phenomena in 2D semiconductor alloys have been extensively studied. In the single-layer group IV transitional metal dichalcogenides (TMDs), it has been proved that there are two structures with very different properties, which are called H phase and T phase, respectively, corresponding to semiconductivity and metallicity [28] (Figure 6.4a). Moreover, these two phases may correspond to more various electronic properties in different materials. It is precisely because of the simultaneous presence of two phases in a single layer of material that the realization of phase transitions becomes possible. There are many ways to induce a phase change. Wherein, alloying may facilitate phase change by reducing the relative energy between the H and T′ phase difference, showing a potential applied on the phase change problem.

Karel-Alexander Duerloo et al. studied the phase transition problem in Mo$_{1-x}$W$_x$Te$_2$ monolayer alloy [31]. The structural phase transition of such temperature-dependent materials in 2D semiconductor alloys has great research significance in the field of PCM devices. 2D semiconductor alloys offer great flexibility, not only in terms of structure and volume but also more importantly

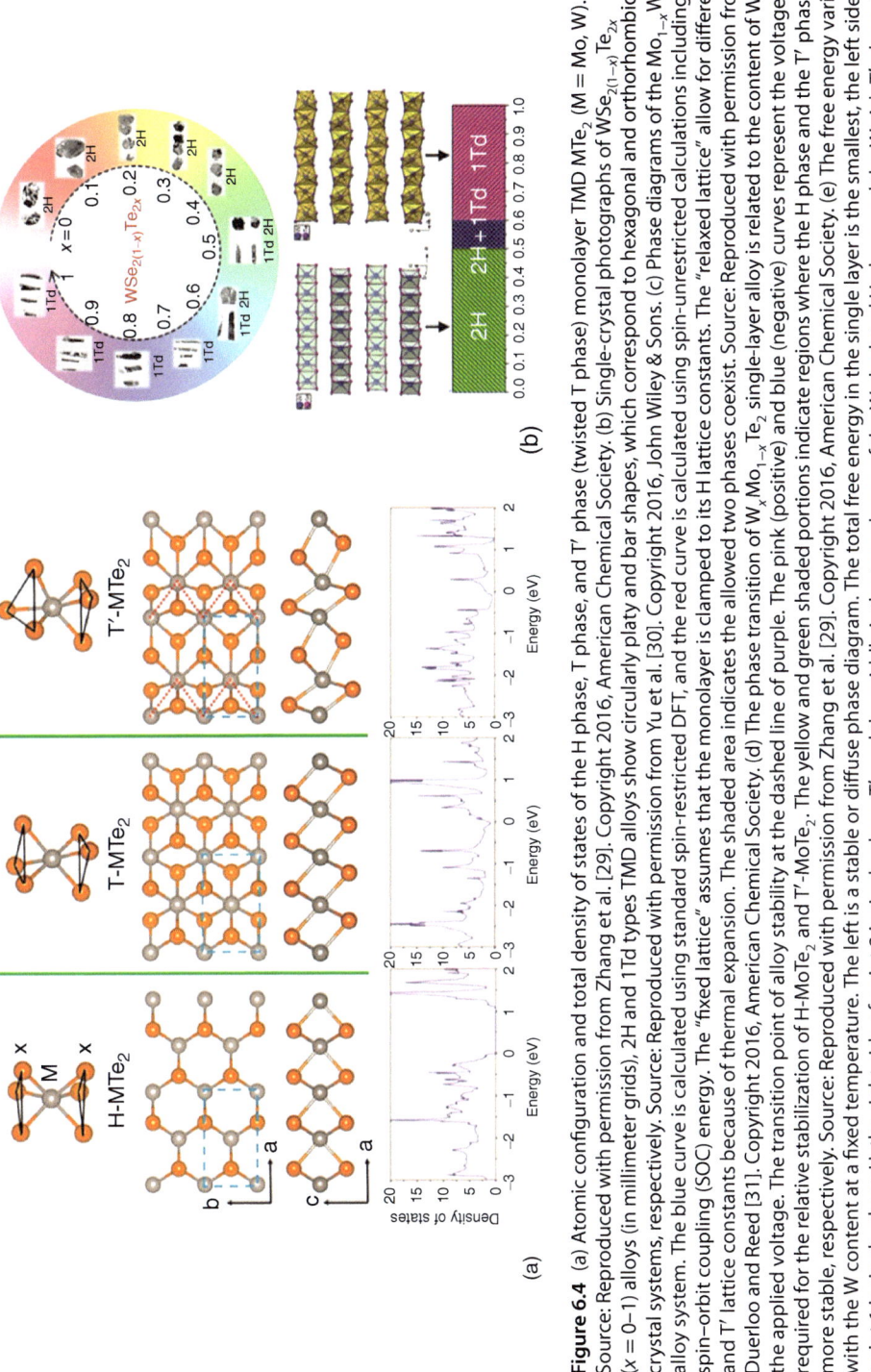

Figure 6.4 (a) Atomic configuration and total density of states of the H phase, T phase, and T′ phase (twisted T phase) monolayer TMD MTe_2 (M = Mo, W). Source: Reproduced with permission from Zhang et al. [29]. Copyright 2016, American Chemical Society. (b) Single-crystal photographs of $WSe_{2(1-x)}Te_{2x}$ (x = 0–1) alloys (in millimeter grids), 2H and 1Td types TMD alloys show circularly platy and bar shapes, which correspond to hexagonal and orthorhombic crystal systems, respectively. Source: Reproduced with permission from Yu et al. [30]. Copyright 2016, John Wiley & Sons. (c) Phase diagrams of the $Mo_{1-x}W_xTe_2$ alloy system. The blue curve is calculated using standard spin-restricted DFT, and the red curve is calculated using spin-unrestricted calculations including spin–orbit coupling (SOC) energy. The "fixed lattice" assumes that the monolayer is clamped to its H lattice constants. The "relaxed lattice" allow for different H and T′ lattice constants because of thermal expansion. The shaded area indicates the allowed two phases coexist. Source: Reproduced with permission from Duerloo and Reed [31]. Copyright 2016, American Chemical Society. (d) The phase transition of $W_xMo_{1-x}Te_2$ single-layer alloy is related to the content of W and the applied voltage. The transition point of alloy stability at the dashed line of purple. The pink (positive) and blue (negative) curves represent the voltages required for the relative stabilization of H-$MoTe_2$ and T′-$MoTe_2$. The yellow and green shaded portions indicate regions where the H phase and the T′ phase are more stable, respectively. Source: Reproduced with permission from Zhang et al. [29]. Copyright 2016, American Chemical Society. (e) The free energy varies with the W content at a fixed temperature. The left is a stable or diffuse phase diagram. The total free energy in the single layer is the smallest, the left side of point 1 is single phase H, the right side of point 3 is single phase T′, and the middle is the two phases of the W-depleted H phase and the W-rich T′ phase. The right side is a metastable or nondiffusion phase diagram. Point 2 represents the transition between the metastable single-phase H and the metastable single-phase T′. Source: Reproduced with permission from Duerloo and Reed [31]. Copyright 2016, American Chemical Society.

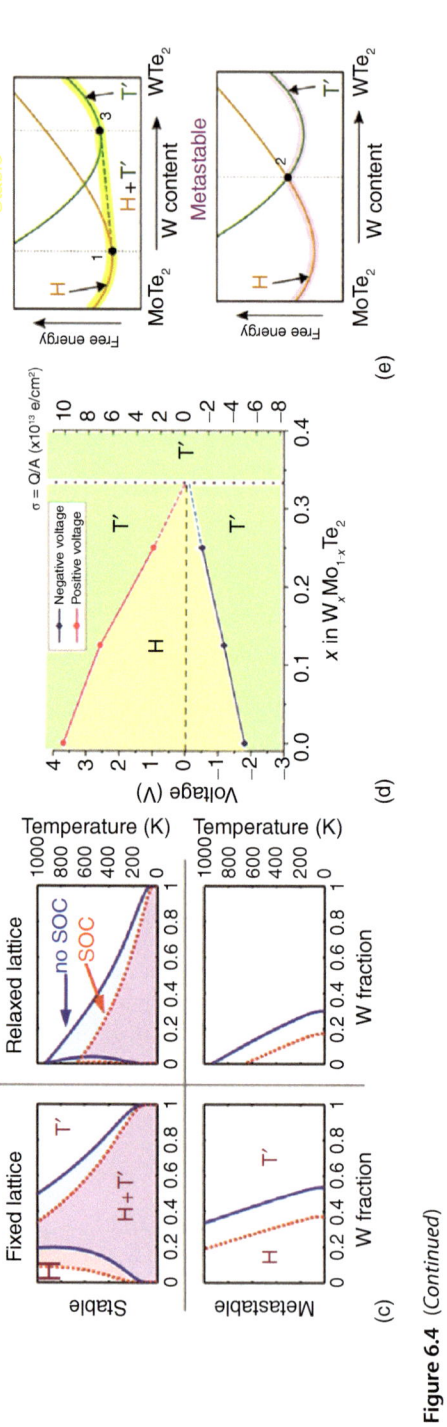

Figure 6.4 (*Continued*)

in terms of electrical and optical properties. Studies on $Mo_{1-x}W_xTe_2$ monolayer alloys have shown that by adjusting the chemical composition of the alloy, the transition temperature (H–T′ phase transition temperatures) in the phase diagram can be varied widely (Figure 6.4e). This allows the phase diagram to be used to derive the growth and thermal processes of a particular phase (for example, Figure 6.4c predicts which phase or mixture phase is thermodynamically dominant at a given temperature and alloy composition). Moreover, the monolayer alloy material can better reduce energy consumption from reducing latent heat and reducing volume.

Later, Chenxi Zhang et al. used DFT theoretical calculations to study the phase stability problem in $W_xMo_{1-x}Te_2$ single-layer alloys [29]. In addition to the effect of W concentration in the alloy, charge regulation is also a viable means of inducing phase transitions. The results show that increasing the W concentration in the $W_xMo_{1-x}Te_2$ single-layer alloy can make the T′ phase tend to be stable, and a phase transition (H phase–T′ phase) occurs at $x = 0.333$ (Figure 6.4d). Before the content of W is less than 0.333, the applied voltage can still undergo a phase change, but the threshold voltage decreases as W increases until the phase is stabilized after $x = 0.333$.

Similarly, the study of $WSe_{2(1-x)}Te_{2x}$ alloys involves a transition between two different structures (semiconductor 2H phase and metal 1Td phase) [30]. The phase of the alloy is also different by adjusting the concentration of Te (Figure 6.4b). When $x = 0$–0.4, $WSe_{2(1-x)}Te_{2x}$ alloy is 2H phase, when $x = 0.5$–0.6, $WSe_{2(1-x)}Te_{2x}$ alloy coexists with 2H phase and 1Td phase, when $x = 0.7 - 1.0$, and the alloy is a 1Td phase. That is, in the complete composition of $WSe_{2(1-x)}Te_{2x}$ alloys, three regions of 2H phase, 2H, and 1Td phases coexist and 1Td phase.

6.2.4 Application of 2D Semiconductor Alloys in the Field of Magnetism

There is no doubt that 2D semiconductor alloys have enormous application potential in the fields of optics and electricity. Even 2D semiconductor alloys have great research value in the field of magnetism. By doping the magnetic atoms into the crystal lattice of the 2D material, neither the structure of the 2D material itself nor the magnetic signal can be introduced.

Sining Dong et al. used the method of Mn doping $SnSe_2$ to grow $Sn_{1-x}Mn_xSe_2$ alloy with higher Mn concentration by molecular beam epitaxy (MBE) method [32]. The concentration of Mn can be adjusted depending on the temperature of the Mn effusion cell. The resulting $Sn_{1-x}Mn_xSe_2$ alloy exhibited significant weak ferromagnetic behavior at room temperature.

Agnieszka Kuc et al. studied the monolayer of MoS_2 ternary alloy with W, Nb, and Se atoms from first-principles calculation. The results show that the electronic structure of ternary alloys has changed significantly compared with MoS_2 monolayer. The calculation of spin splitting energies of materials with different compositions also shows that alloys can adjust the optical and spintronic properties of materials in a wide range, which is of great significance to the application of two-dimensional materials in the field of spintronics [11].

6.2.5 Improve Device Performance

Two-dimensional materials have great application prospects, but 2D materials themselves still have great limitations. 2D alloy materials are the expansion and optimization of 2D materials. By doping, alloying, etc., some properties that the 2D material itself has cannot be realized. Or the improvement of the original properties can be achieved by alloying; for example, the improvement of electrical and photoelectric properties can be achieved by adjusting the energy band structure. The change in carrier concentration, the increase in carrier mobility, the increase in anisotropy, and the suppression of electron–hole recombination [33] can be achieved by alloying.

For example, Figure 6.5a,b is the optical absorption of the $SnSe_{2(1-x)}S_{2x}$ alloy [17]. With the increase of S content, in the visible spectral range, the optical absorption is significantly enhanced. Moreover, the comparison of the optical

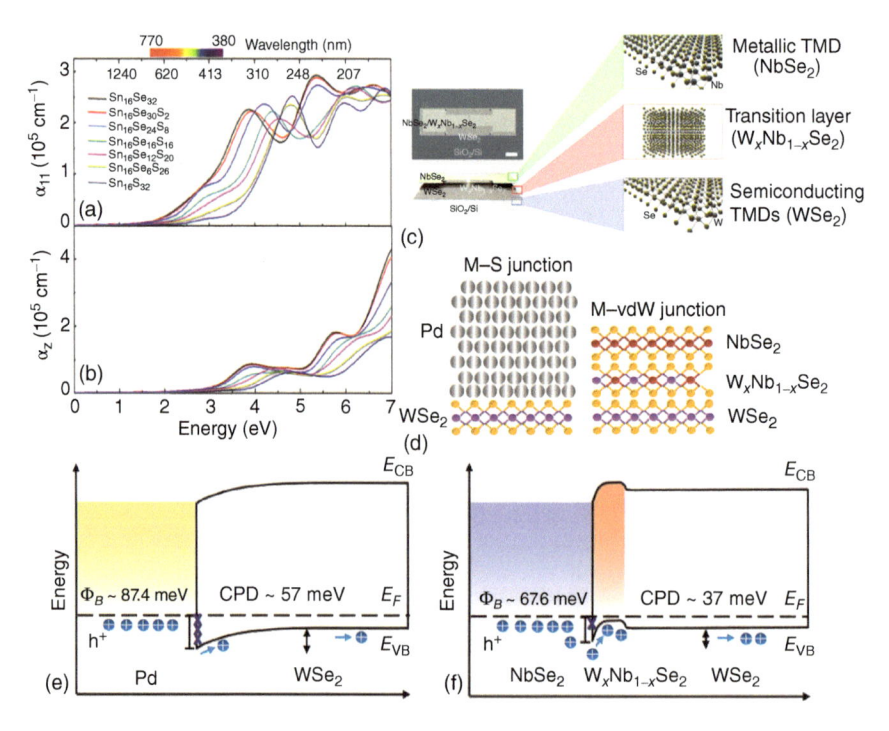

Figure 6.5 Optical absorption coefficients along the (a) xy-plane and (b) z-directions of the $SnSe_{2(1-x)}S_{2x}$ alloys with different S compositions. Source: Reproduced with permission from Huang et al. [17]. Copyright 2016, American Chemical Society. (c) Optical image and schematic drawing of WSe_2-based bottom-gate FET with $NbSe_2$ electrode. Source: Reproduced with permission from Kim et al. [34]. Copyright 2016, American Chemical Society. (d) Schematic diagram of the atomic structure of metal Pd-semiconductor WSe_2 (M–S) junction and $NbSe_2/W_xNb_{1-x}Se_2/WSe_2$ (M–vdW) junction comprising an alloy transition layer. The schematic energy band diagrams of (e) metal Pd-semiconductor WSe_2 channel and (f) $NbSe_2/W_xNb_{1-x}Se_2/WSe_2$ van der Waals (M–vdW) junction with Schottky contact. CPD, contact potential difference. Source: Reproduced with permission from Kim et al. [35]. Copyright 2016, American Chemical Society.

absorption of the xy-plane and the z-axis indicates a strong anisotropy. The improvement of optical absorption of materials by alloying has great significance in optical applications.

Building a semiconductor heterostructure is a method of designing devices commonly used in 2D materials in the optical/electrical field. However, a high contact resistance is generated between the three-dimensional metal electrode and the two-dimensional semiconductor.

In order to solve the metal–semiconductor contact problem, Ah Ra Kim et al. added a ternary alloy $W_xNb_{1-x}Se_2$ layer as a buffer layer (Figure 6.5c) between the semiconductor WSe_2 and the metal $NbSe_2$ [34]. The composition of $W_xNb_{1-x}Se_2$ can be precisely controlled by plasma-enhanced atomic layer deposition (PEALD). The electrical properties of $W_xNb_{1-x}Se_2$ alloys with different compositions will be significantly different, which will have a greater impact on the charge transport in the device. In general, this approach can greatly reduce the height of the Schottky barrier, resulting in lower contact resistance, further improving the performance of the field effect transistor (FET) device. A new idea for building devices and improving device performance is proposed, which also shows that 2D semiconductor alloys have great application potential. Subsequently, they studied the origin of this introduction of the alloy buffer layer resulting in low contact resistance [35]. Compared to conventional metal–junction (M–S junction) devices (left side of Figure 6.5d), M–vdW (van der Waals) junction devices with $W_xNb_{1-x}Se_2$ alloy transition layer (on the right side of Figure 6.5d) have a longer hot carrier lifetime. The main reason for this effect is a significant reduction in the number of surface defects on the material and a reduction in the Schottky barrier. The reduction of the Schottky barrier is also due to the cleanliness of the 2D material interface. Different from the Fermi level pinning effect caused by a large number of interface traps in the M–S junction, the M–vdW junction interface has only a few dangling bonds, which can greatly reduce the Schottky barrier height (Figure 6.5e,f). At the same time, because of the existence of the transition layer, the tunnel width of the quantum tunneling is also significantly reduced, and the tunneling efficiency is also greatly improved. The results show that the $NbSe_2/W_xNb_{1-x}Se_2/WSe_2$ heterojunction with mixed transition layer has remarkable advantages. This proves that 2D semiconductor alloys also have potential application value in optimizing device structure.

Moreover, for the $SnSe_{2(1-x)}S_{2x}$ alloy, the carrier mobility of the nanosheet can be adjusted between SnS_2 and $SnSe_2$ by controlling the composition of the alloy [18]. Moreover, under the illumination, the mobility of carriers is greatly improved. Sijie Liu et al. obtained an n-type $Nb_{0.125}Re_{0.875}Se_2$ alloy by alloying p-type $ReSe_2$ [27] and obtained a strong photodetection capability.

The expansion and improvement of material properties of 2D semiconductor alloys provide a means to improve device performance. As more and more alloys are synthesized, the performance will get better and better. It will also provide an increasingly good material base for the emergence of high-performance devices.

6.3 Preparation Methods for 2D Alloys

Some two-dimensional semiconductor alloys have been shown to be thermodynamically stable, which theoretically demonstrates the feasibility of experimental preparation. There are many ways to synthesize two-dimensional semiconductor alloys, including chemical vapor deposition (CVD), physical vapor deposition (PVD), chemical vapor transport (CVT), and MBE. As with 2D materials, layer control is also very important for 2D alloys, especially for single-layer materials. CVD, PVD, and MBE can directly prepare single-layer alloys, and CVT can also control the number of layers of two-dimensional alloys by mechanical peeling.

6.3.1 Chemical Vapor Transport (CVT)

CVT is a common method for the conventional preparation of single crystals and is now widely used in the preparation of 2D materials. Also in the preparation of 2D semiconductor alloys, CVT is still a very important method.

In the general CVT preparation process, the raw materials and the transport agent (such as I_2) need to be sealed in a vacuum quartz tube. Then, it is placed in a tube furnace having a temperature gradient (for example, a dual-temperature zone tube furnace), and the material becomes a gas phase in a high-temperature region and then crystallizes at a low-temperature end to form a single crystal. For single-crystal materials prepared by general CVT, it is necessary to obtain sheets of different layers or even single layers by mechanical exfoliation (Figure 6.6a) [36]. The combination of CVT preparation materials and mechanical exfoliation methods is a good thin layer sample preparation method for 2D materials and 2D alloys.

For example, CVT method is used to prepare $WSe_{2(1-x)}Te_{2x}$ ($x = 0–1$) alloys (Figure 6.6b) [30]. First, a certain stoichiometric ratio of W, Te, Se, and iodine

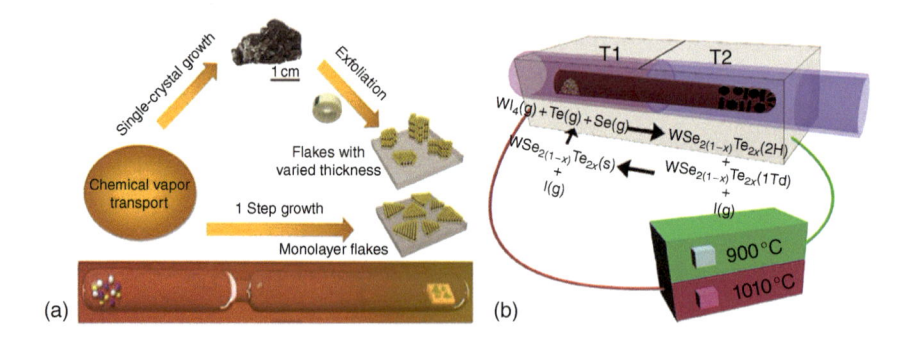

Figure 6.6 (a) Schematic diagram of the growth process of the CVT method. The top is a conventional two-step single-crystal growth + mechanical stripping method to obtain sheets of different thicknesses. The bottom is a CVT directly grown monolayer film. Source: Reproduced with permission from Hu et al. [36]. Copyright 2017, John Wiley & Sons. (b) Schematic diagram of the synthesis and morphology of $WSe_{2(1-x)}Te_{2x}$ ($x = 0–1$) alloy single crystal. Source: Reproduced with permission from Yu et al. [30]. Copyright 2016, John Wiley & Sons.

as a transport agent was sealed in a vacuum quartz tube with a vacuum of 10^{-6} Torr. The reaction zone was set to 850 °C and the growth zone was set to 900 °C for pretreatment, wherein this temperature setting prevented the transport of the sample. The reaction zone was then set to 1010 °C and the growth zone was set to 900 °C for three days with a temperature gradient to provide single-crystal growth. The furnace was naturally cooled to room temperature. Finally, a $WSe_{2(1-x)}Te_{2x}$ single-crystal material can be obtained in the growth zone. Nanosheets of different layers can be obtained by mechanical exfoliation. Similarly, in the CVT synthesis process of $ReS_{2(1-x)}Se_{2x}$, the Re, S, and Se elements are sealed into the quartz tube in a certain ratio, and a certain amount of Br_2 is placed as a transporting agent. The vacuum quartz tube was placed in a tube furnace and grown at 1030 °C for 10 days [16, 34]. It is generally common to use CVT growth to provide a certain temperature gradient using two temperature zones of a dual-temperature zone tube furnace and then to obtain a high-quality single-crystal block under such a temperature gradient for a certain period of time. Therefore, we can get a lot of 2D semiconductor alloys, in addition to the above, including $TiS_{3(1-x)}Se_{3x}$ [37], $SnSe_{2(1-x)}S_{2x}$ [18], $Nb_{(1-x)}Ti_xS_3$ [38], and so on.

6.3.2 Physical Vapor Deposition (PVD)

The process of preparing a 2D alloy by physical vapor deposition (PVD) is almost identical to the process of preparing a 2D material. Probably, the process is to evaporate the raw materials under certain conditions and then deposit them on the substrate in the low-temperature zone to prepare the corresponding materials. The following will be introduced from the specific experimental process.

Q. Feng et al. used a three-zone tube furnace to grow $MoS_{2(1-x)}Se_{2x}$ ($x = 0$–0.40) monolayer alloys and $MoS_{2(1-x)}Se_{2x}$ ($x = 0.41$–1.00) monolayer alloys (Figure 6.7a) [22, 40]. During the growth of the $MoS_{2(1-x)}Se_{2x}$ ($x = 0$–0.40) alloy, the $MoSe_2$ powder and the MoS_2 powder were placed in the first- and second-temperature zones, respectively. During the evaporation process, the two-temperature zones (T_1 and T_2) were above 940 °C (typically $T = 940$–975). The grown substrate was a SiO_2/Si substrate with an oxide thickness of approximately 270 nm and placed in a third-temperature zone that was not heated. The $MoS_{2(1-x)}Se_{2x}$ alloy was suitable for growth at temperatures of 600–700 °C (Figure 6.7a, blue dotted line). The growth process needs to be assisted by a carrier gas, which is a mixture of Ar and H_2. By adjusting a series of conditions during the growth process, such as evaporation temperature, deposition gradient temperature (best at 50 °C/cm), and carrier gas H_2 flow rate (H_2 is preferably 0.5 sccm when Ar is 2 sccm), the quality of growth has been improved. In the growth of $MoS_{2(1-x)}Se_{2x}$ ($x = 0.41$–1.00) single-layer alloy, in addition to the $MoSe_2$ and MoS_2 powders in the first- and second-temperature zones, the temperature in front of the first-temperature zone was about 300 °C. Place the selenium powder at the location (here mainly to provide selenium-enriched environment). The first- and second-temperature zones had a temperature of about 950–965 °C and a deposition temperature of about 600–770 °C. The best carrier gases were Ar (2 sccm) and H_2 (1 sccm). The growth conditions, such

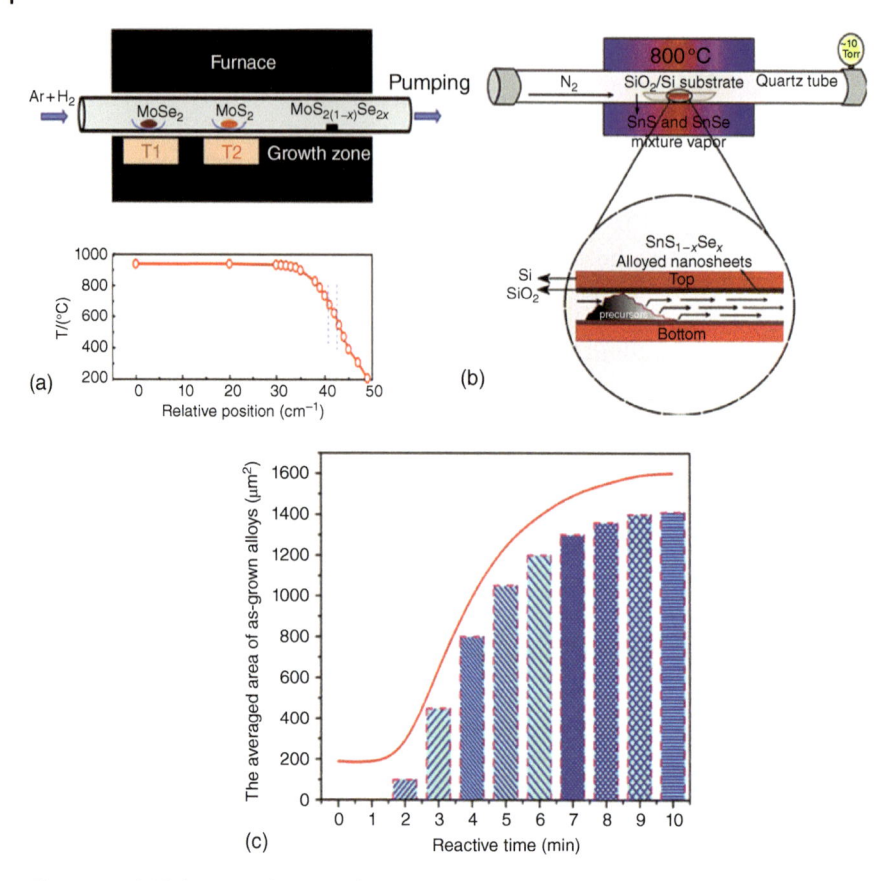

Figure 6.7 (a) Schematic diagram of a three-zone tube furnace for growing a MoS$_{2(1-x)}$Se$_{2x}$ monolayer alloy and a temperature profile associated with the position. Source: Reproduced with permission from Feng et al. [22]. Copyright 2014, John Wiley & Sons. (b) Schematic diagram of NGPVD-grown SnS$_{1-x}$Se$_x$ alloy. The lower part is a magnified image of the black ellipse showing the detailed orientation of the growth of the alloy. (c) During the growth of SnS$_{1-x}$Se$_x$, the histogram of the average area of the grown sample is taken as a function of reaction time. Source: Reproduced with permission from Gao et al. [39]. Copyright 2009, Royal Society of Chemistry.

as evaporation temperature, deposition temperature gradient, and deposition temperature, large domain size and edge orientation are adjusted to obtain high-quality crystals. Comparing the two growth processes, the process was very similar, and the obtained crystals tend to be the same material, but there were still many differences in the growth process, including the need for selenium-rich conditions in the latter process and the pressure during the growth process, growth time, etc.

Wei Gao et al. used the so-called narrow gap physical vapor deposition (NGPVD) method to grow large ultrathin SnS$_{1-x}$Se$_x$ ($0 \leq x \leq 1$) alloy nanosheets [39]. A homogeneous mixture of SnS and SnSe powder was placed at one end of the SiO$_2$/Si substrate and then covered with another SiO$_2$/Si substrate(Figure 6.7b). Place this part in the center of the quartz tube but do not put it in the furnace first. Before heating, it is necessary to pass N_2 into the quartz tube to remove oxygen and adjust the pressure to 10 Torr by a vacuum pump. The furnace was then heated to 800 °C, pushed into a quartz tube, and placed in the center of the heated zone. It was heated at different times to get different samples. Immediately after the growth is completed, the substrate is pulled out for cooling. Figure 6.7c shows the relationship between the average area of the sample and the growth time during the growth process.

6.3.3 Chemical Vapor Deposition (CVD)

The CVD method is one of the most common methods for growing 2D materials and is one of the most commonly used methods for growing 2D semiconductor alloys. The specific steps of CVD in the growth of 2D semiconductor alloys are described below in several practical works.

A CVD growth of a $ReS_{2x}Se_{2(1-x)}$ monolayer alloy is taken as an example. Figure 6.8a is a schematic diagram of CVD synthesis of $ReS_{2x}Se_{2(1-x)}$ alloy and its atomic structure [15]. Using a home-made CVD system, ReO_3 was placed as a precursor for growth, and sulfur powder and selenium powder were used as the S source and the Se source, respectively. An alloy-grown mica plate was placed on a ReO_3 ceramic boat, and a layer of molecular sieve was coated on the ReO_3 powder to control the uniform growth of the film. Adjusting the ratio of the S source to the Se source can result in alloys of different compositions. The temperatures of the S source and the Se source were 200 and 300 °C, respectively. Ar gas was introduced as a carrier gas and grown at a temperature of 700 °C for five minutes.

Sima Umrao et al. synthesized a $MoS_{2(1-x)}Se_{2x}$ single-layer alloy by low-pressure CVD [41]. Using a two-heating-zone furnace (Figure 6.8b), the S powder and the Se powder were placed in the center of furnace 1 and MoO_3 was placed in the center of furnace 2. The grown substrate is placed face down on the MoO_3 powder and arranged in a specific position (this position is related to the gaseous MoO_3 concentration). The evaporation of S and Se can be controlled by furnace 1 and furnace 2 synthesizes the $MoS_{2(1-x)}Se_{2x}$ alloy at 800 °C. Different layers of alloys can be obtained by varying the concentration and temperature of MoO_3. Moreover, alloys grown at different locations have different forms.

Xidong Duan et al. used a home-built CVD system (Figure 6.8c) to place the WS_2 and WSe_2 powders at different locations (i.e. different temperatures) in the tube furnace and place the grown SiO_2/Si substrate at the lower end of the system [4]. The tube furnace has a fixed temperature gradient. The vapor pressure and ratio of WS_2 and WSe_2 are precisely controlled by directly changing the temperature of WS_2 and the temperature of the position-dependent WSe_2. Further, a $WS_{2x}Se_{2-2x}$ alloy nanosheet having a fully adjustable chemical composition was obtained.

As one of the most commonly used methods for preparing 2D materials, CVD has a wide range of applications in the preparation of 2D alloys. The 2D semiconductor alloy that has been successfully fabricated by CVD has $HfS_{2(1-x)}Se_{2x}$ [20], $CdS_xSe_{(1-x)}$ [19], $Mo_{1-x}W_xS_2$ [42, 43], $MoS_{2x}Te_{2(1-x)}$ [44], $MoS_{2(1-x)}Se_{2x}$

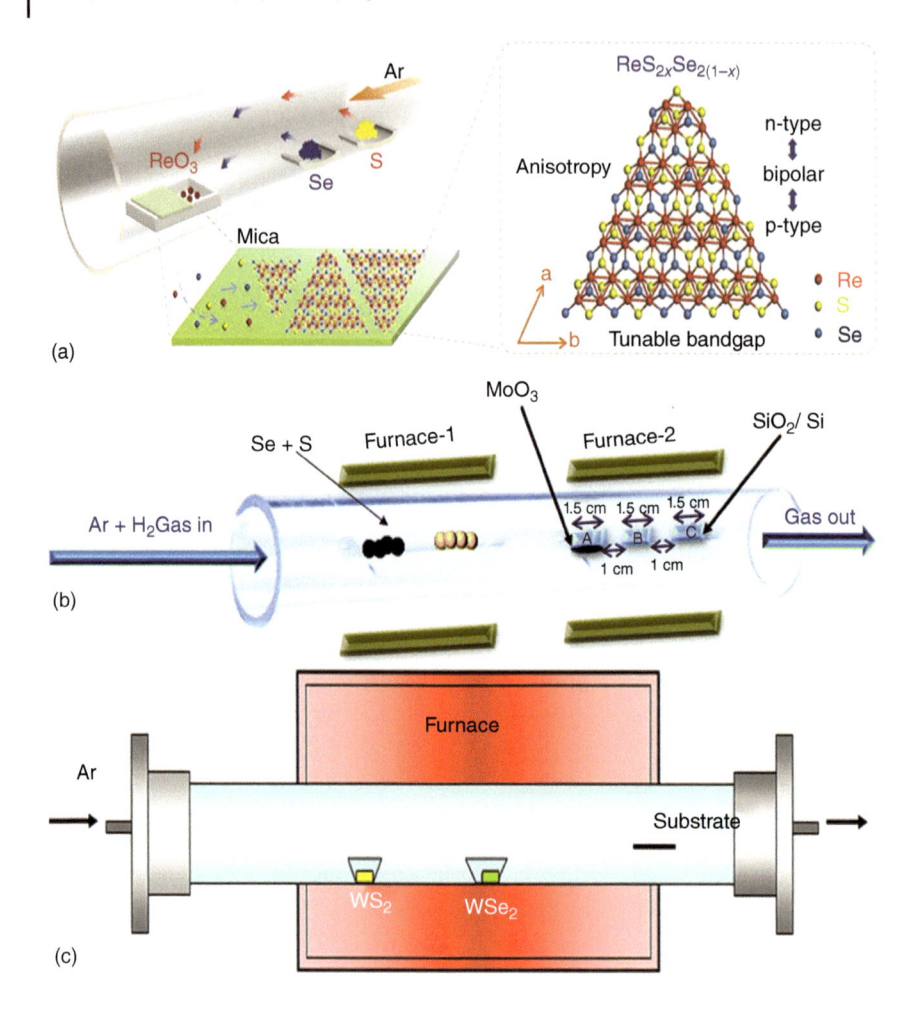

Figure 6.8 (a) Schematic diagram of CVD synthesis of ReS$_{2x}$Se$_{2(1-x)}$ alloy and its atomic structure. Source: Reproduced with permission from Cui et al. [15]. Copyright 2017, John Wiley & Sons. (b) Schematic diagram of two-heating-zone furnace for synthesizing MoS$_{2(1-x)}$Se$_{2x}$ alloys. Source: Reproduced with permission from Umrao et al. [41]. Copyright 2009, Royal Society of Chemistry. (c) Schematic diagram of CVD system for WS$_{2x}$Se$_{2-2x}$ alloy growth. Source: Reproduced with permission from Duan et al. [4]. Copyright 2016, American Chemical Society.

[3, 25, 45], MoS$_{2(1-x)}$Se$_{2x}$ [46], and the like. Further, some fixed component alloys are also synthesized by CVD, such as Co$_{0.16}$Mo$_{0.84}$S$_2$ alloy [47]. It is conceivable that more and more excellent alloys will be prepared in the future.

6.4 Characterizations of 2D Alloys

Several common alloying characterization methods include atomic resolution scanning transmission electron microscopy (STEM), Raman spectroscopy, photoluminescence (PL) spectroscopy, and X-ray photoelectron spectroscopy

(XPS). STEM can directly distinguish individual atoms, get the arrangement of atoms, and then analyze their electronic structure. Raman spectroscopy is an important tool for structural analysis of materials.

6.4.1 STEM

STEM can be used to directly observe the distribution of atoms. The lattice structure is analyzed based on the arrangement of the atoms. STEM is the most commonly used means of characterizing lattice structures. When analyzing the alloy, you can use the light and darkness and size of the STEM image to directly observe the approximate structure or supplemented by other means, such as X-ray diffraction (XRD), energy disperse spectroscopy (EDS), etc., to obtain more detailed and accurate results.

For example, in the characterization of the $WSe_{2(1-x)}Te_{2x}$ monolayer alloy, the arrangement of the alloy and the alloy phase at different Te concentrations can be obtained according to the STEM display of the atomic arrangement (Figure 6.9f–i) [49]. In the STEM characterization of the $Mo_{1-x}W_xS_2$ single-layer alloy, in order to more clearly distinguish Mo and W atoms, atomic resolved electron energy loss spectroscopy (EELS) analysis and quantitative simulation of annular dark field (ADF) curves were performed [48] (Figure 6.9a–e).

As the elements that make up the alloy become more diverse, it is important to determine the atomic arrangement of the various alloys [50]. The electronic structure of the responsive material is then obtained. Therefore, in order to better characterize the atomic arrangement, STEM and a combination of various techniques and methods provide a good idea.

6.4.2 Raman Spectroscopy

Raman spectroscopy is a technique for studying the scattered light of the interaction of molecules with light. The molecular structure of the material can be inferred from the obtained spectral peak information (such as wave number, shape, intensity, etc.). Each substance has its own characteristic Raman spectrum. Raman spectral analysis has been widely used in 2D materials. For alloy materials, the frequency shift is generally used to determine the composition of the alloy, and the peak broadening is used to indicate the alloying degree [51]. Next, the characterization of 2D semiconductor alloys by Raman spectroscopy is mainly analyzed by using TMDCs as an example.

It is known that the characteristic Raman peaks of TMDC materials are A_{1g} and $E_{2g'}$ (expressed as A_1' and E' in a single layer), for example, the A_1' peak of monolayer of MoS_2 is about 400 cm^{-1}, and the E' peak is about 384 cm^{-1}. In the monolayer 2D semiconductor alloys, the Raman spectrum of the material also changes as the composition of the alloy changes. For different materials, there may be different Raman vibration modes and different behaviors, for example, one-mode behavior and two-mode behavior [2]. The former means that there is only one peak of the same Raman mode in the two extreme cases of alloy change ($x = 0$ and $x = 1$). The latter refers to the fact that the same Raman mode is a separate peak in the alloy during the composition change from $x = 0$ to $x = 1$.

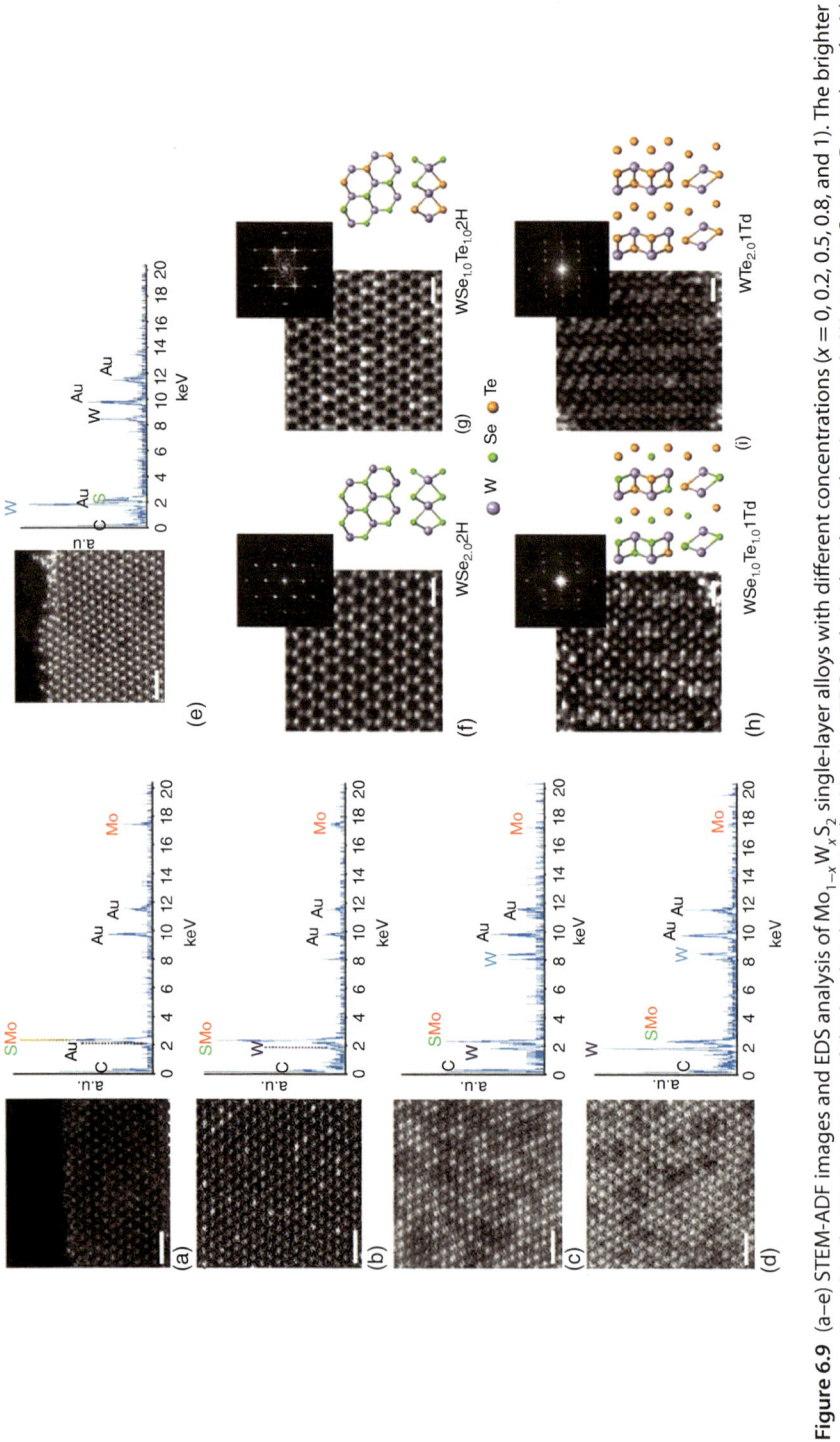

Figure 6.9 (a–e) STEM-ADF images and EDS analysis of $Mo_{1-x}W_xS_2$ single-layer alloys with different concentrations (x = 0, 0.2, 0.5, 0.8, and 1). The brighter spots correspond to the W atom sites and the less bright ones to the Mo sites. EDS analysis also showed an increase in W content. Source: Reproduced with permission from Dumcenco et al. [48]. Copyright 2016, Springer Nature. (f–i) Atomic resolution STEM characterization of $WSe_{2(1-x)}Te_{2x}$ monolayer alloys with different Te concentrations. Different atoms can be distinguished by image intensity. The picture shows the STEM test image, the fast Fourier transform (FFT) image, and the atomic model. Source: Reproduced with permission from Yu et al. [30]. Copyright 2016, John Wiley & Sons.

For example, Raman spectroscopy of a $Mo_{1-x}W_xS_2$ single-layer alloy is an example [51, 52]. It has been observed in experiments that there are two first-order Raman patterns of activity in single-layer MoS_2 and WS_2: A_1' and E' [21]. The atomic displacements of the two modes are shown in Figure 6.10b. A_1' is an out-of-plane vibration containing only S atoms, and E' is an in-plane vibration including a Mo atom (or a W atom) and an S atom. For the $Mo_{1-x}W_xS_2$

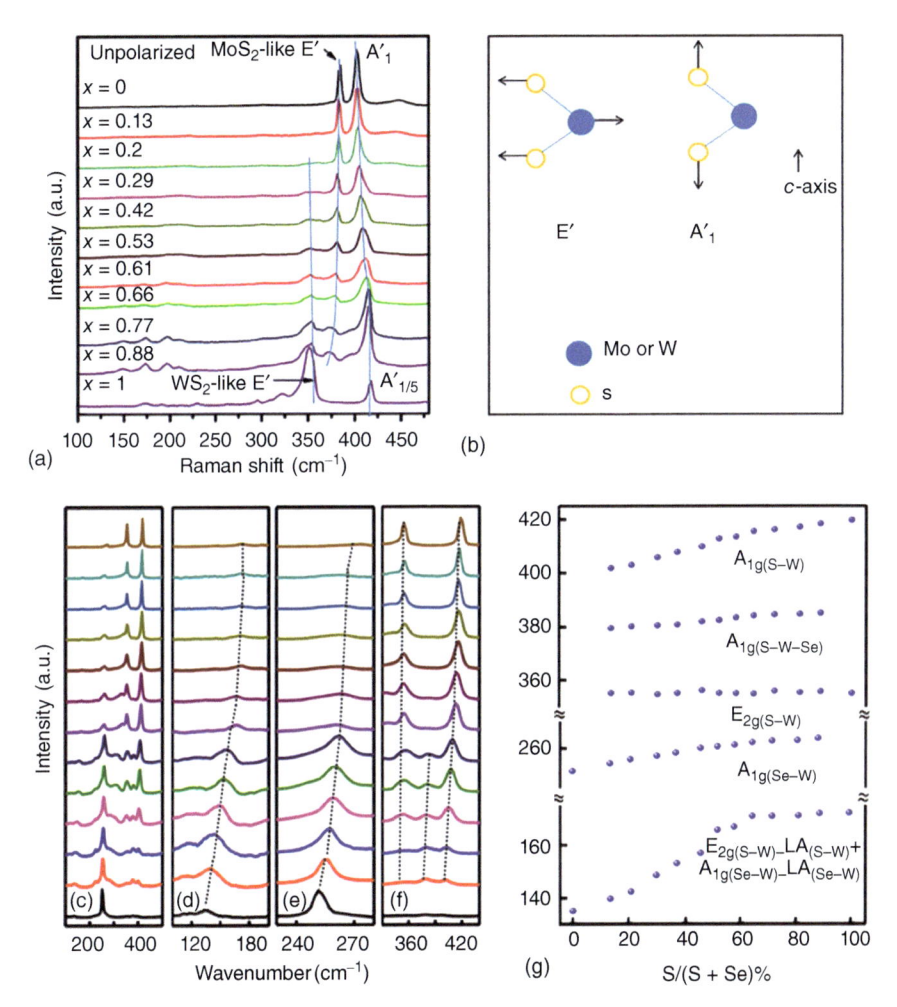

Figure 6.10 (a) Raman spectra of $Mo_{1-x}W_xS_2$ monolayer alloys with different W contents. The three blue lines indicate the frequency shift of the E' and A_1' peaks as the W content x differs. (b) Schematic diagram of the atomic shift of the Raman-active E' and $A1'$ modes in $Mo_{1-x}W_xS_2$ monolayer alloys. Source: Reproduced with permission from Chen et al. [51]. Copyright 2009, Royal Society of Chemistry. (c–g) The evolution of Raman spectroscopy with components in $WS_{2x}Se_{2-2x}$ single-layer alloy nanosheets. (c) Full-range Raman spectra; (d) $E_{2g(S-W)} - LA_{(S-W)} + A_{1g(Se-W)} - LA_{(Se-W)}$ mode of the $WS_{2x}Se_{2-2x}$ nanosheets; (e) A_{1g} of Se–W mode; and (f) E_{2g} of S–W mode, A_{1g} of S–W–Se mode, and A_{1g} of S–W. (g) Raman spectra peak position shifts with increasing S atomic ratio for the five Raman modes. Source: Reproduced with permission from Duan et al. [4]. Copyright 2016, American Chemical Society.

Figure 6.11 (a) The PL spectrum of 2H-phase $WSe_{2(1-x)}Te_{2x}$ ($x = 0-0.6$) monolayer alloys. (b) The dependence of the bandgap of the $WSe_{2(1-x)}Te_{2x}$ ($x = 0-1.0$) single-layer alloy on the content of Te component. At $x = 0.5-0.6$, a phase transition from the semiconductor (2H phase) to the metal (1Td phase) occurs. Source: Reproduced with permission from Yu et al. [30]. Copyright 2016, John Wiley & Sons. (c, e) The PL spectra associated with the bulk and single-layer $ReS_{2(1-x)}Se_{2x}$ alloys. (d, f) Bandgap and composition relationship of bulk and single-layer $ReS_{2(1-x)}Se_{2x}$ alloys. Source: Reproduced with permission from Wen et al. [16]. Copyright 2017, John Wiley & Sons.

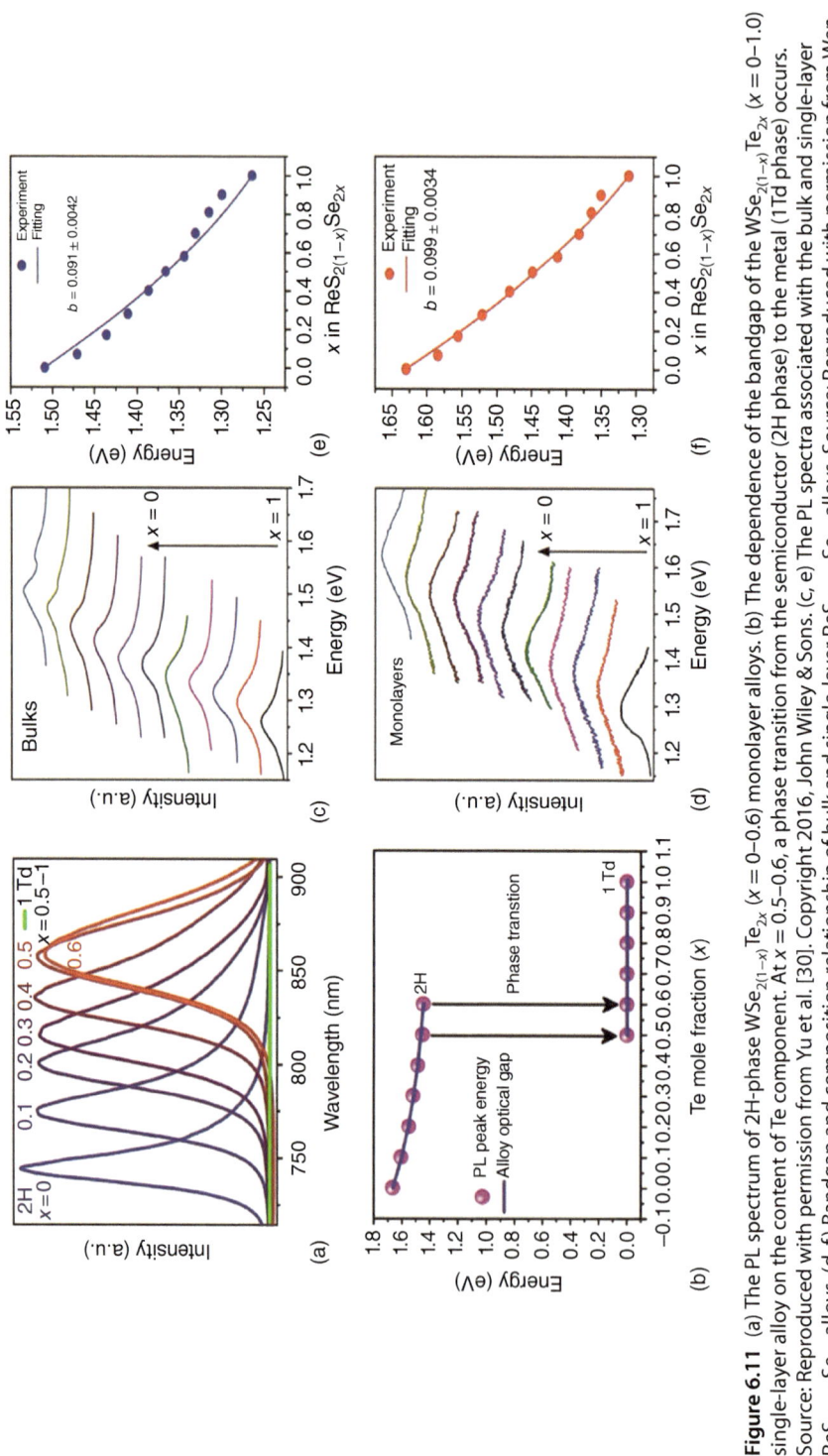

single-layer alloy, A_1' and E' exhibit single-mode behavior and dual-mode behavior, respectively (Figure 6.10a). Specifically, the A_1' mode continuously changes with the difference of the component X of W. The E' mode shows two phonon branches associated with MoS_2 and WS_2. As the composition of W increases, the intensity of the E' mode associated with WS_2 increases, while the intensity of the E' mode associated with MoS_2 decreases. The intensity common to the A_1' and E' modes is increasing because of resonance enhancement.

In the Raman spectroscopy of $WS_{2x}Se_{2-2x}$ alloy nanosheets, five modes were shown, namely $A_{1g(S-W)}$ mode, $A_{1g(Se-W)}$ mode, $A_{1g(S-W-Se)}$ mode, $E_{2g(S-W)}$ mode, and $E_{2g(S-W)} - LA_{(S-W)} + A_{1g(Se-W)} - LA_{(Se-W)}$ mode [4]. As the composition changes, significant peak position shifts and relative intensity evolution can be seen (Figure 6.10c–g). The expected evolution of structure and composition in the nanosheets was further confirmed.

6.4.3 Photoluminescence (PL) Spectrum

PL is a commonly used means of characterizing the optical bandgap of semiconductor materials. For 2D semiconductor alloys, bandgap tuning is one of its most important properties; in order to characterize the regulation of the bandgap, PL spectrum is a good way.

Figure 6.11a is a PL spectrum excited by a 532 nm laser of a $WSe_{2(1-x)}Te_{2x}$ single-layer alloy [30]. All 2H-phase $WSe_{2(1-x)}Te_{2x}$ single-layer alloys exhibited emission bands with peaks ranging from 744 nm (pure WSe_2) to 857 nm (near-infrared). For the alloy material of the 1Td phase ($x \geq 0.6$), no PL signal appears. A continuous change in the bandgap from 1.67 to 1.44 eV can be obtained from the obtained PL spectrum (Figure 6.11b). And when x = 0.5–0.6, a phase transition from a semiconductor (2H phase) to a metal (1Td phase) occurs.

For the PL measurement of the $ReS_{2(1-x)}Se_{2x}$ alloy, the bandgap variation associated with the composition can also be obtained [16]. In the bulk $ReS_{2(1-x)}Se_{2x}$ alloy, as the Se content increases (x varies from 0 to 1), the PL emission energy changes from 1.52 to 1.26 eV (Figure 6.11c,e). In the single-layer $ReS_{2(1-x)}Se_{2x}$ alloy, the PL emission energy changes from 1.62 to 1.31 eV (Figure 6.11d,f).

6.5 Doping of 2D Semiconductors

The need for tunable bandgap and the need for more performance expansion or optimization of optoelectronic devices have contributed to the rapid development of 2D alloys. Similar to the manner of alloying, the manner in which impurity atoms are introduced into the material to alter the intrinsic properties of the 2D material is also a way to achieve more functionalization of the 2D material. In fact, when the content of a certain component is relatively small, the 2D alloy can be regarded as atomic doping. Although there are few impurity atoms incorporated, it has a great influence on the properties of the material and may even affect the inherent properties of the material.

For example, for chemical doping of MoS_2, the electronic band structure of MoS_2 can be altered by introducing different atoms. In the regulation of the physical properties of semiconductors, doping and alloying have similarities. By

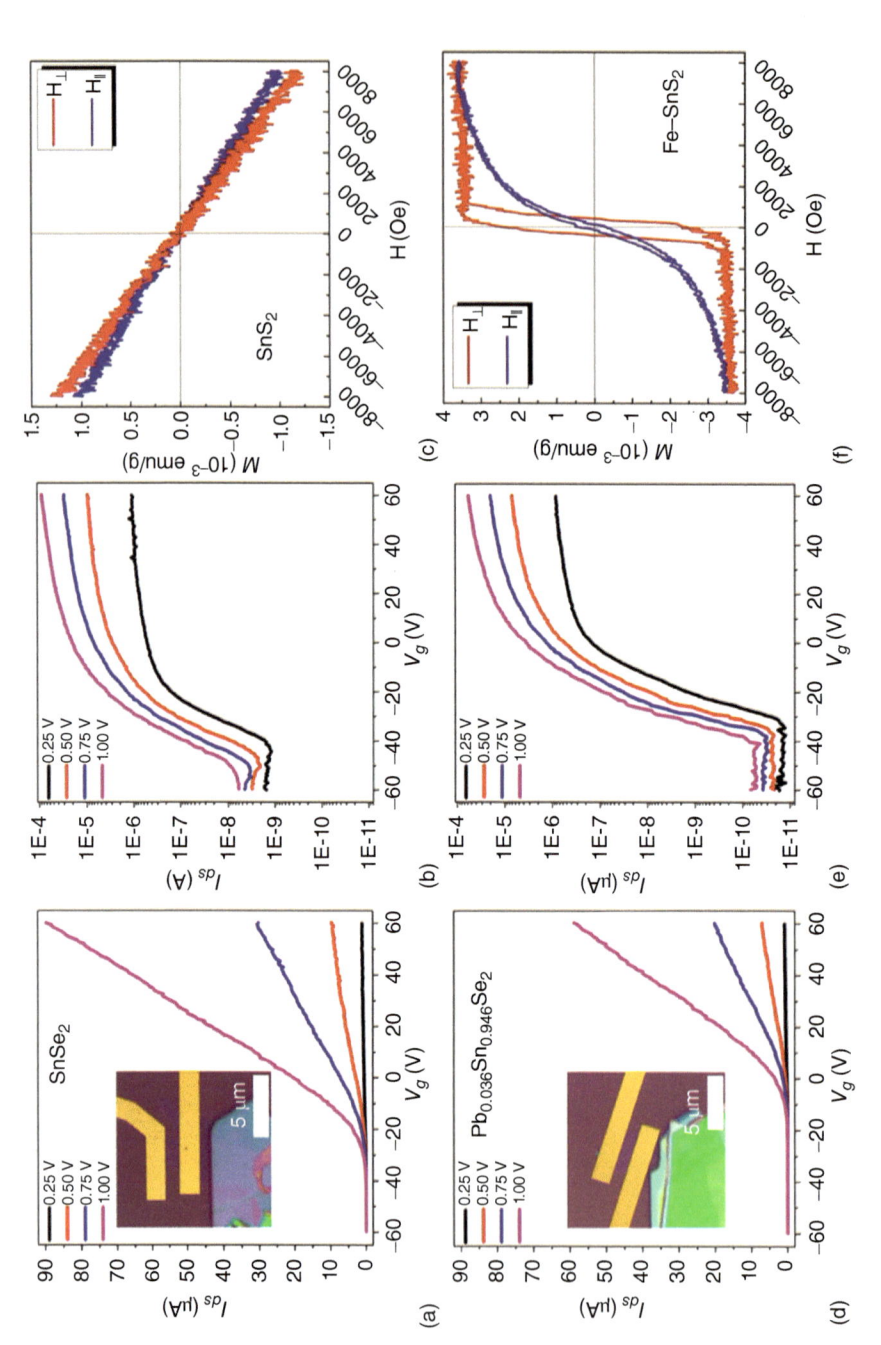

Figure 6.12 (a, b, d, e) Comparison of the electrical transport properties of $SnSe_2$ and $Pb_{0.036}Sn_{0.964}Se_2$ nanosheets. (a, b) Transfer characteristics of the $SnSe_2$ single-layer devices of linear and logarithmic scales, respectively. The inset is an optical image of the $SnSe_2$ device. (d, e) Transfer characteristics of $Pb_{0.036}Sn_{0.964}Se_2$ single-layer devices with linear and logarithmic scales, respectively. Source: Reproduced with permission from Liu et al. [58]. Copyright 1990, IOP Publishing. (c, f) Magnetic hysteresis loops measured by a VSM (vibrating sample magnetometer) at 2 K for SnS_2 and $Fe_{0.021}Sn_{0.979}S_2$, respectively. Source: Li et al. 2017 [59]. https://www.nature.com/articles/s41467-017-02077-z. Licensed under CC BY 4.0.

doping impurity atoms, it is possible to control the conductivity type, carrier concentration, conductivity, bandgap and so on. In addition to the regulation of electron transport properties, the incorporation of magnetic atoms (Fe, Co, Ni, Mn, etc.) can also be magnetically controlled or converted, which is the origin of 2D dilute magnetic semiconductors (2D-DMS) [53–57].

Doping is an important tool for controlling the optical, electrical, and magnetic properties of 2D semiconductor materials. Various 2D semiconductor alloys with tunable bandgap (e.g. $Mo_{1-x}W_xS_2$ [21], $WS_{2x}Se_{2-2x}$ [4], etc.) fully demonstrate the importance of doping in the regulation of optical and electrical properties. For example, Pb doping of a single layer of $SnSe_2$ [58], XPS measurement gives a doping concentration of about 3.6% ($Pb_{0.036}Sn_{0.964}Se_2$). TEM shows that the Pb atoms are uniformly doped. Electrical measurements show that the switching ratio of the $Pb_{0.036}Sn_{0.964}Se_2$ single-layer FETs is 2 orders of magnitude larger than that of the $SnSe_2$ single layer and higher than any other measurement (Figure 6.12a,b,d,e). This indicates that doping is a very effective means to adjust the properties of 2D semiconductor materials and has great application prospects in the fields of electronics and optoelectronics. In addition, doping is also a potentially viable means of implementing exciton devices. The preparation of uniformly doped 2D semiconductors is the basis for exciton manipulation. For example, in the Sb-doped MoS_2 crystal, doping results in the formation of impurity levels, and the presence of impurity levels in the bandgap causes excitons to change [60].

Not only in potential applications in optics/electrics, the use of magnetic 2D materials in spintronics has received increasing attention. For example, in the Fe-doped SnS_2 single layer [59], the FET measurement shows excellent photoelectric properties, and the magnetic measurement changes magnetically. Antiferromagnetic SnS_2 exhibits ferromagnetic behavior after incorporation of about 2.1% Fe atoms ($Fe_{0.021}Sn_{0.979}S_2$) (Figure 6.12c,f).

References

1 Ji, Q., Zhang, Y., Zhang, Y., and Liu, Z. (2015). *Chem. Soc. Rev.* 44: 2587.
2 Xie, L.M. (2015). *Nanoscale* 7: 18392.
3 Li, H., Duan, X., Wu, X. et al. (2014). *J. Am. Chem. Soc.* 136: 3756.
4 Duan, X., Wang, C., Fan, Z. et al. (2016). *Nano Lett.* 16: 264.
5 Wang, H., Yuan, H., Sae Hong, S. et al. (2015). *Chem. Soc. Rev.* 44: 2664.
6 Ci, L., Song, L., Jin, C. et al. (2010). *Nat. Mater.* 9: 430.
7 Yuge, K. (2009). *Phys. Rev. B* 79: 144109.
8 Komsa, H.P. and Krasheninnikov, A.V. (2012). *J. Phys. Chem. Lett.* 3: 3652.
9 Ge, C.H., Li, H.L., Zhu, X.L., and Pan, A.L. (2017). *Chin. Phys. B* 26: 034208.
10 Kutana, A., Penev, E.S., and Yakobson, B.I. (2014). *Nanoscale* 6: 5820.
11 Kuc, A. and Heine, T. (2015). *Electronics* 5: 1.
12 Xi, J., Zhao, T., Wang, D., and Shuai, Z. (2014). *J. Phys. Chem. Lett.* 5: 285.
13 Cai, H., Chen, B., Blei, M. et al. (2018). *Nat. Commun.* 9: 1927.
14 Zhao, N., Zhu, Y.F., and Jiang, Q. (2018). *J. Mater. Chem. C* 6: 2854.
15 Cui, F., Feng, Q., Hong, J. et al. (2017). *Adv. Mater.* 29: 1705015.

16 Wen, W., Zhu, Y., Liu, X. et al. (2017). *Small* 13: 1603788.
17 Huang, Y., Chen, X., Zhou, D. et al. (2016). *J. Phys. Chem. C* 120: 5839.
18 Wang, Y., Le Huang, L.H., Li, B. et al. (2017). *J. Mater. Chem. C* 5: 84.
19 Xia, J., Zhao, Y.X., Wang, L. et al. (2017). *Nanoscale* 9: 13786.
20 Wang, D., Zhang, X., Guo, G. et al. (2018). *Adv. Mater.* 30: 1803285.
21 Chen, Y.F., Xi, J.Y., Dumcenco, D.O. et al. (2013). *ACS Nano.* 7: 4610.
22 Feng, Q., Zhu, Y., Hong, J. et al. (2014). *Adv. Mater.* 26: 2648.
23 Feng, Q.L., Mao, N.N., Wu, J.X. et al. (2015). *ACS Nano.* 9: 7450.
24 Im, H.S., Myung, Y., Park, K. et al. (2014). *RSC Adv.* 4: 15695.
25 Mann, J., Ma, Q., Odenthal, P.M. et al. (2014). *Adv. Mater.* 26: 1399.
26 Li, X., Lin, M.W., Basile, L. et al. (2016). *Adv. Mater.* 28: 8240.
27 Liu, S., Huang, L., Wu, K. et al. (2016). *Appl. Phys. Lett.* 109: 112102.
28 Chhowalla, M., Shin, H.S., Eda, G. et al. (2013). *Nat. Chem.* 5: 263.
29 Zhang, C., Kc, S., Nie, Y. et al. (2016). *ACS Nano.* 10: 7370.
30 Yu, P., Lin, J., Sun, L. et al. (2017). *Adv. Mater.* 29: 1603991.
31 Duerloo, K.A. and Reed, E.J. (2016). *ACS Nano.* 10: 289.
32 Dong, S., Liu, X., Li, X. et al. (2016). *APL Mater.* 4: 032601.
33 Zhao, S., Wu, J., Jin, K. et al. (2018). *Adv. Funct. Mater.* 28: 1802011.
34 Kim, A.R., Kim, Y., Nam, J. et al. (2016). *Nano Lett.* 16: 1890.
35 Kim, Y., Kim, A.R., Yang, J.H. et al. (2016). *Nano Lett.* 16: 5928.
36 Hu, D., Xu, G., Xing, L. et al. (2017). *Angew. Chem. Int. Ed. Engl.* 56: 3611.
37 Agarwal, A., Qin, Y., Chen, B. et al. (2018). *Nanoscale* 10: 15654.
38 Yang, S., Wu, M., Shen, W. et al. (2019). *ACS Appl. Mater. Interfaces* 11: 3342.
39 Gao, W., Li, Y., Guo, J. et al. (2018). *Nanoscale* 10: 8787.
40 Ho, C.H., Huang, Y.S., Liao, P.C., and Tiong, K.K. (1999). *J. Phys. Chem. Solids* 60: 1797.
41 Umrao, S., Jeon, J., Jeon, S.M. et al. (2017). *Nanoscale* 9: 594.
42 Park, J., Kim, M.S., Park, B. et al. (2018). *ACS Nano.* 12: 6301.
43 Yang, K., Wang, X., Li, H. et al. (2017). *Nanoscale* 9: 5102.
44 Yin, G., Zhu, D., Lv, D. et al. (2018). *Nanotechnology* 29: 145603.
45 Gong, Y., Liu, Z., Lupini, A.R. et al. (2014). *Nano Lett.* 14: 442.
46 Li, H., Zhang, Q., Duan, X. et al. (2015). *J. Am. Chem. Soc.* 137: 5284.
47 Li, B., Huang, L., Zhong, M. et al. (2015). *ACS Nano.* 9: 1257.
48 Dumcenco, D.O., Kobayashi, H., Liu, Z. et al. (2013). *Nat. Commun.* 4: 1351.
49 Freyland, W., Zell, C.A., El Abedin, S.Z., and Endres, F. (2003). *Electrochim. Acta* 48: 3053.
50 Azizi, A., Antonius, G., Regan, E. et al. (2019). *Nano Lett.* 19: 1782.
51 Chen, Y., Dumcenco, D.O., Zhu, Y. et al. (2014). *Nanoscale* 6: 2833.
52 Song, J.G., Ryu, G.H., Lee, S.J. et al. (2015). *Nat. Commun.* 6: 7817.
53 Cheng, Y.C., Zhu, Z.Y., Mi, W.B. et al. (2013). *Phys. Rev. B* 87: 100401.
54 Mishra, R., Zhou, W., Pennycook, S.J. et al. (2013). *Phys. Rev. B* 88: 144409.
55 Ramasubramaniam, A. and Naveh, D. (2013). *Phys. Rev. B* 87: 195201.
56 Seixas, L., Carvalho, A., and Neto, A.H.C. (2015). *Phys. Rev. B* 91: 155138.
57 Sun, L.L., Zhou, W., Liang, Y.H. et al. (2016). *Comput. Mater. Sci.* 117: 489.
58 Liu, J., Zhong, M., Liu, X. et al. (2018). *Nanotechnology* 29: 474002.
59 Li, B., Xing, T., Zhong, M. et al. (2017). *Nat. Commun.* 8: 1958.
60 Zhong, M., Shen, C., Huang, L. et al. (2019). *npj 2D Mater. Appl.* 3: 1.

7

Properties of 2D Heterostructures

7.1 Conception and Categories of 2D Heterostructures

The development of grapheme [1] pushes the experiments and theory on 2D materials. Meanwhile, structural instability and excellent advantages were excavated in several systems on 2D materials. The quantum effect is more distinct in low dimensionality, which exhibited several excellent electrical [2–4], thermionic [5], mechanical [6], and optical properties. Grapheme is an ultrahigh mobility 2D material that leads to various applications, but its gaplessness limited the popularity in the application of electronics.

The 2D material can be classified into the following categories on the basis of different structures of natural bandgap: insulators, conductors, topological insulators [7, 8], and super conductors [9]. Nevertheless, there are Xenes [10, 11], MX_2 [12–15], nitrides [16, 17], oxides [18, 19], MXenes [20, 21], and a special structure as the essential kinds of 2D materials according to the type of element. Xenes contain only one single element, such as grapheme [1], black phosphorus [22], borophene [23], silicone [24], germanene [25], stanene [26], arsenene [27], and antimonene [28]. MX_2 also named transition metal dichalcogenide (TMD) possesses a sandwich crystal structure, in which an M atom is in the middle of two layers X atoms in the vertical structure. Generally, the bandgap in the majority of MX_2 ranges from 1.0 to 2.5 eV, which is utilized as high-performance field-effect transistors and photodetectors. As a representative, MoS_2 became the focus attributed to the high mobility and moderate bandgap. Nitrides include hexagonal boron nitride (h-BN) [16], Carbon-doped boron nitride (CBN) [29], AlN [30], g-C_3N_4 [31], GaN [32], etc., most of which are insulators playing an important role in the heterostructures. Oxides refer to hydroxides, and MXenes are composed of carbon or nitrogen with metal. Some particular structures were also designed in 2D materials. For example, the Janus monolayers of MoSSe that could break out-of-plane structural symmetry have been successfully synthesized [33]. Heterostrctures based on abundant 2D materials have large potential for applications in the future.

Recently, heterostructures as derivatives of 2D materials have come into the researchers' vision. 2D materials possess a special stratified structure in which atoms are bonded by covalent bonds within atom layers and by weak van der Waals forces between atom layers. Nevertheless, all of the atoms of traditional semiconductors are combined by covalent bonds. The approach of distinguishing 2D materials from bulk materials is: two-dimensional material atoms are bonded by van der Waals force and bulk material atoms are bonded by covalent. The 2D materials always possess the few layers meeting the ultrathin and subminiature requirements in the contemporary applications. Heterostructures comprise two or more different components and could be classified as 0D–2D, 1D–2D, 2D–2D, and 3D–2D patterns. However, 2D–2D is the most meaningful pattern in the majority of circumstances. We focus on heterostructures with 2D–2D pattern in which both components are 2D materials in this section.

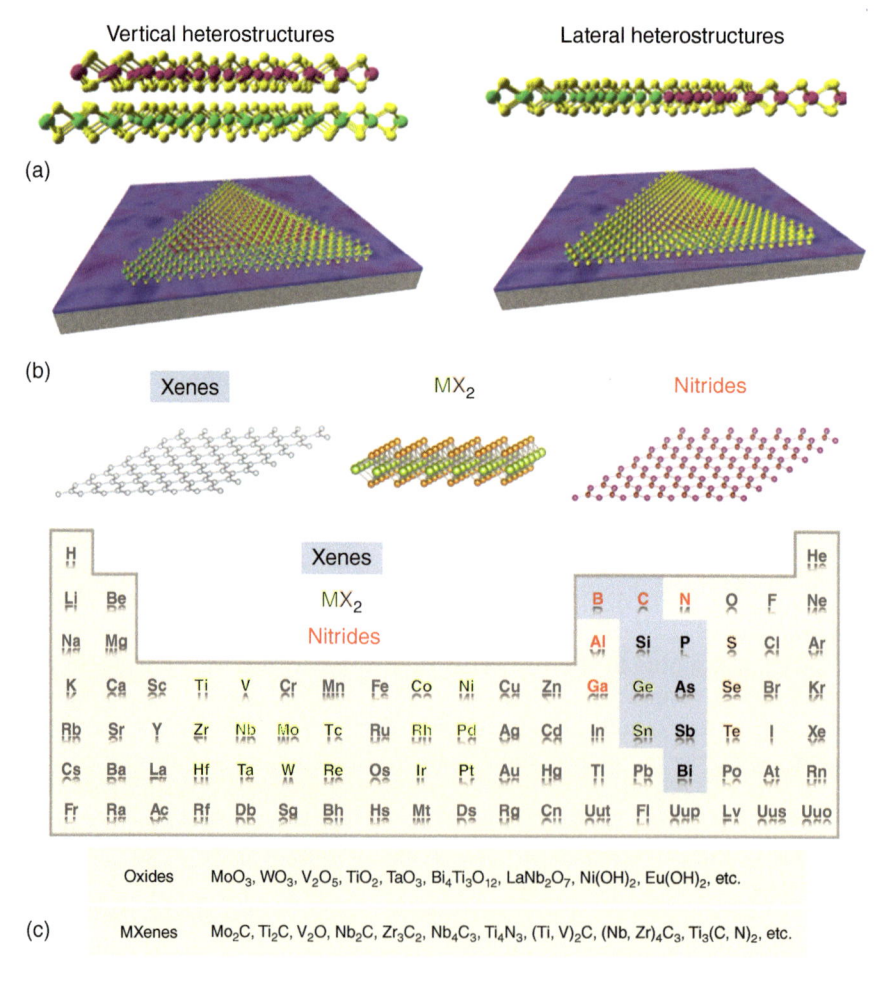

Figure 7.1 (a) Vertical and lateral heterostructures of 2D materials. Source: Reprinted with permission from Duesberg et al. [56]. Copyright 2014, Nature Publishing Group. (b) Atomic structures of the main three types of 2D materials. Source: Cui et al. 2018 [57]. Reproduced with permission of Springer Nature. (c) Elements that constitute Xenes, MX_2, and nitrides are marked by different signs on the periodic table, and 2D oxides and MXenes are individually listed in the lower table. Source: Reprinted with permission from Cui et al. [57]. Copyright 2017, Nature Publishing Group.

2D heterostructures stack by multiple kinds of 2D materials to realize particular function and could be classified as vertical and lateral heterostructure by the location of electrode; therefore, the way of electron transport would be different. In vertical heterostructures, top electrode stacks on top of the 2D heterostructures and bottom electrode locates under the 2D heterostructures; hence, electrons would transport along the vertical direction. In lateral heterostructures, electrodes and 2D heterostructures stitching by matching components are in the same plane; hence, carriers would transport along the lateral direction. The differences are shown in Figure 7.1. In vertical heterostructures, the free clean surface of 2D materials without any dangling bands depleted of the results of lattice mismatch. Moreover, a large reserve of vertical heterostructures exists in various optional 2D materials, thus providing a flexible strategy for the construction of specific heterostructures. In lateral heterostructures, covalent bonding is considered as the general bonding among the 2D materials. Thus, vertical, lateral, and other heterojunctions are not differentiated and are collectively referred to as 2D heterostructures. According to the function of materials, there are several types as follows. A 2D semiconductor with a semiconductor is common, and there is also a 2D semiconductor with an insulator, a semiconductor with semimetal, an insulator with semimetal, etc. Because of the multifarious, excellent properties in 2D materials, the combination of 2D materials makes more senses, which constitute more particular structures to realize specific applications.

7.2 Advantages and Application of 2D Heterostructures

Semiconductor heterostructures are the basic units in electronic and optoelectronic devices such as high-mobility transistors, lasers, and light-emitting diodes. Conventional semiconductor heterostructures primarily consist of III–V or II–VI compound semiconductor materials by covalent bonding. Because of atomic diffusion during epitaxial growth, an ideal abrupt doping profile cannot be formed. In addition, conventional epitaxial growth of heterostructures requires matching of the lattice constants of the constituent materials; therefore, the choice of semiconductor heterostructure material composition is limited. The interaction force between the two-dimensional materials is van der Waals force, and there is no dangling bond. Different two-dimensional materials are assembled into two-dimensional heterostructures by van der Waals force in vertical heterostructures. There is no atomic diffusion between different two-dimensional materials, which can form an ideal mutated-junction. Because there is no limitation of lattice matching, any two-dimensional material can be combined into a vertical heterojunction according to the specific application requirements, which provides a broad platform for designing high-performance semiconductor devices. Otherwise, the interlayer quantum coupling effect in vertical heterojunctions can lead to novel physical properties, and the electrical and optical properties of the device can be modulated by adjusting the heterostructure interface. The use of two-dimensional heterostructures has now enabled multiple functions in electronic and optoelectronic device applications.

a) *Electronic device*: Different two-dimensional materials have different functions and play an irreplaceable role in heterojunctions. In the vertical heterojunction, the graphene/h-BN vertical heterojunction mainly uses h-BN to reduce the charge trap between the insulating layer and graphene and enhance the mobility of carriers in graphene; transition-metal dichalcogenide (TMDC)/graphene vertical heterojunction mainly combines the photoresponse of TMDCs and the high conductivity of graphene in high-performance photoresponsive devices; TMDC/TMDC vertical heterojunctions mainly combine different energy band structures of two materials to control carrier transport and facilitate storage and high-performance photoresponse.

The tunnel diode is a two-dimensional heterojunction with a representative application. The tunnel diode can obtain differential negative resistance and is very useful for nonlinear components. Roy et al. used a discrete gate to modulate the MoS_2/WSe_2 heterojunction to implement a tunnel diode and observe differential negative resistance at low temperatures (below 175 K) [34]. In addition, its structural diagram and electrical properties are shown in Figure 7.2, and significant negative differential resistance can be seen. In the tunnel device, the carrier is injected in the form of band-to-band tunneling (BTBT), which can achieve lower subthreshold swing in the field-effect transistor [35]. In addition to the tunneling diode, the two-dimensional heterojunction can also constitute a common PN junction diode device. As shown in Figure 7.2b, the Bp/MoS_2 heterojunction diode has better rectification characteristics and gate control capability [36].

b) *Photodetectors*: The built-in electric field formed at the interface of the heterojunction can effectively separate the electron–hole pairs generated by the illumination, so that the photodetector based on the two-dimensional heterojunction has a fast response speed. In addition, the vertical heterostructure minimizes the carrier transport distance, making it possible to achieve fast response. Figure 7.3a,b show an array of WS_2/MoS_2 few-layer heterojunctions with a photoresponse of up to 2.3 A/W at a wavelength of 450 nm [37]. Flexible electrical testing of MoS_2/WS_2 heterojunction arrays demonstrates reasonable signal-to-noise ratio.

Lei Ye et al. have shown that a vertical photogate heterostructure of BP-on-WSe_2 demonstrated near-infrared multiband absorption, high photoresponsivity, and polarization sensitivity [38]. It can be seen from Figure 7.3c,d that there are different light responses at different light wavelengths, and the magnitude of the photocurrent varies under linearly polarized light in different directions. Heterostructures have better dichroism than a single material.

Most single-layer TMDC materials and black phosphorus are direct bandgap semiconductor materials, and their electron–hole radiation recombination efficiency is much higher than that of indirect bandgap semiconductors, making it easier to achieve high-quantum efficiency light-emitting devices. For the first time, Cheng et al. observed electroluminescence in the WSe_2/MoS_2 heterojunction, but because the heterojunction is a staggered heterojunction, it is impossible to effectively limit the injected carriers to the luminescent region, resulting in lower radiation recombination efficiency. Therefore, the

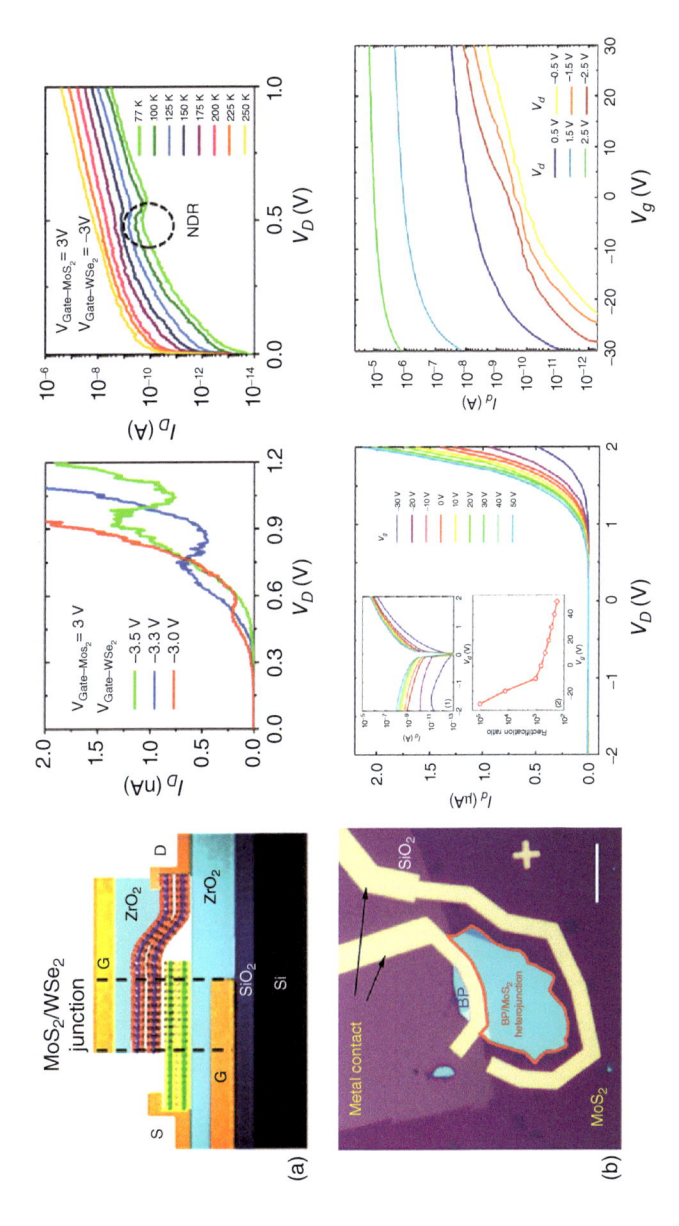

Figure 7.2 (a) Schematic diagram and electrical characteristics of MoS_2/WSe_2 heterojunction structure. Source: Reprinted with permission from Roy et al. [34]. Copyright 2015, American Chemical Society. (b) Bp/MoS_2 heterojunction structure, rectification characteristics, and gate regulation diagram. Source: Reprinted with permission from Deng et al. [36]. Copyright 2014, American Chemical Society.

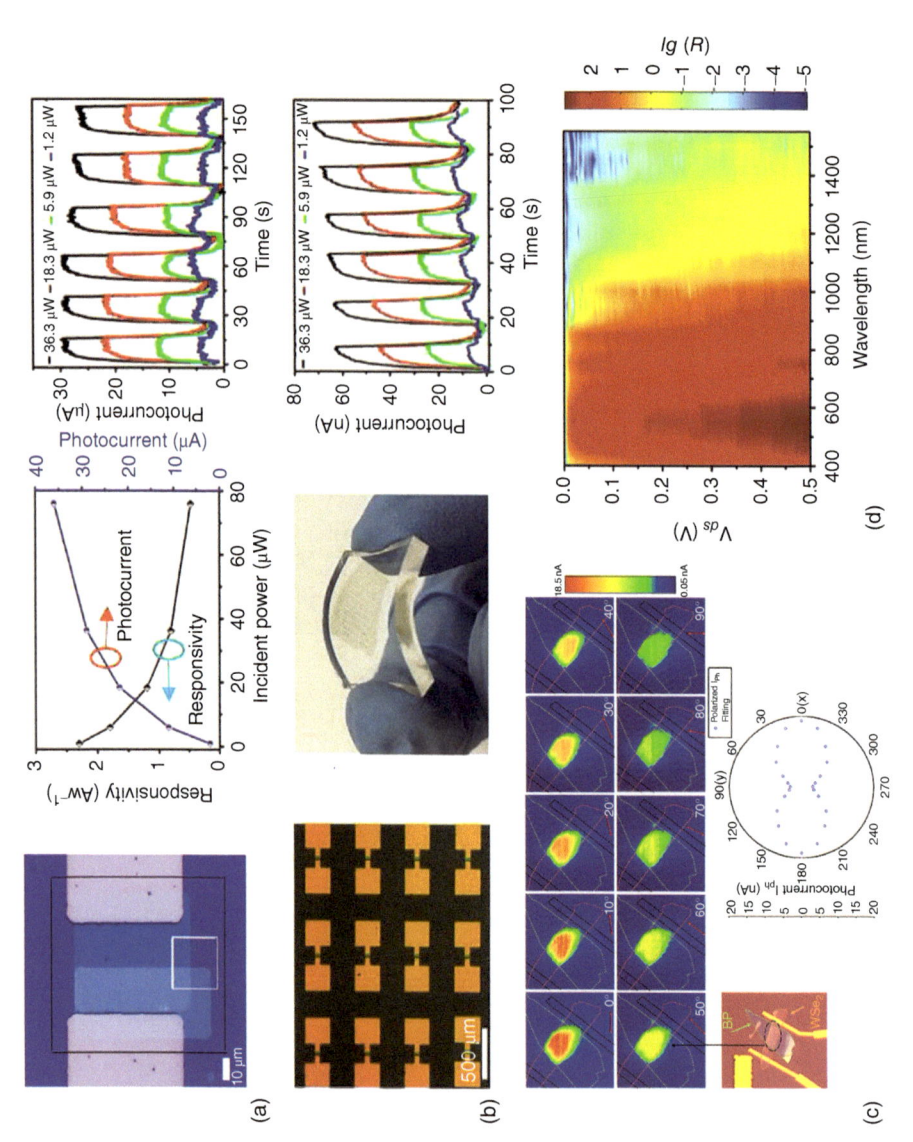

Figure 7.3 (a) WS$_2$/MoS$_2$ heterojunction optical microscopy and photoelectric performance (b) array alignment device and flexible electrical testing. Source: Reprinted with permission from Xue et al. [37]. Copyright 2016, American Chemical Society. (c) Infrared polarized photoresponse of the BP-on-WSe$_2$ photodetectors. (d) The photoresponsivity (R) changes as a function of drain bias (V_{ds}) under different wavelength illumination at 1 mW/cm^2 incident illumination power density. Source: Reprinted with permission from Ye et al. [38]. Copyright 2017, Elsevier.

quantum efficiency is low [39]. In order to effectively confine the injected carriers to the light-emitting region and increase the luminous efficiency, the light-emitting diode generally adopts a quantum well structure. Withers et al. combined a two-dimensional material with different bandgaps into a graphene/h-BN/TMDC/h-BN/graphene heterojunction to form a single quantum well, which achieved an external quantum effect of 1%. By further repeating the structure to form a multiple quantum well, an external quantum efficiency of 8.4% comparable to that of an organic light-emitting diode is achieved [40].

c) *Other devices*: The unique physicochemical properties of the two-dimensional material, as well as the surface-to-volume ratio close to the theoretical maximum and the high surface activity, make it a significant advantage in gas sensing applications. In addition, two-dimensional materials can be conveniently fabricated into field-effect transistors, which have broad prospects in low-power, high-sensitivity gas sensor applications.

In recent years, a number of two-dimensional multi-iron materials, ferroelectric and ferromagnetic materials, have been discovered. In particular, the Curie temperature of some materials has reached room temperature. These two-dimensional materials have been found to be more novel under the condition of extremely thin materials. The physical mechanism has promoted the development of two-dimensional multi-iron, ferroelectric, and ferromagnetic materials to more devices such as memories.

7.3 Preparation Methods for 2D Heterostructures

The method of 2D heterostructures preparation is essential for the device performance. Different growth methods for 2D heterostructures would attain to the diverse results. There are two processes for choosing to find a most appropriate effect: chemical method and mechanical transfer method, which conclude liquid method and dry method, respectively. Firstly, the preparation of 2D flake will be introduced briefly. There are two kinds of methods to obtain the 2D components according to the procedure of crystallization. One is generated by mechanical stripping from the small crystal, and the other obtains is generated from the physical and chemical preparation directly. Different types of crystal blocks are burned according to different growth conditions. The flakes by this way always process good integrity. However, the different lawyers are attached by different forces of Van der Waals. The mechanical stripping of the material with a large van der Waals force between the layers becomes more difficult, especially single layer. Chemical vapor deposition (CVD) and physical vapor deposition methods are usually easier to obtain a single layer, with a problem of material structure imperfections, which would affect the device performance. The mechanical transfer method started a long-time ago and is still extensively utilized in 2D heterostructures today because of its simplicity, convenience, and flexibility. The two mechanical transfer methods would be introduced as follows.

7.3.1 Mechanical Transfer: Liquid Method and Dry Method

As we know, liquid methods are performed within liquid condition and dry methods without the liquid. The characteristics of transfer methods decide the condition to be used. Some 2D materials are unstable in the liquids environment. Therefore dry transfer method has great advantages. However, the intensity of contact in dry method is weaker than that in liquid method. The approaches of both the methods would be introduced as follows.

a) *Liquid method*: As mentioned already, there are several ways to obtain the flakes on silicon, such as mechanical peeling, liquid-phase ultrasound, and lithium ion intercalation. Polymethyl-methacrylate (PMMA) plays an essential role in liquid transfer method. Reducing the wrinkles or distortion on PMMA is beneficial to the transfer process. Generally, PMMA would be evenly coated on the surface of the silicon by spinner. The spinner parameter could be adjusted by specific situations. For example, the thickness could be controlled by the speed of spin. The low speed leads to a thick membrane with the problem of nonuniform covering; the fast speed leads to a thin membrane, which would trigger PMMA breaking and wrinkle during the process of transfer. Therefore, the selection of the speed needs to be decided by transfer. After the coating, drying on the drying station is indispensable because the moisture evaporation from the membrane makes it harder. The temperature and time on the drying station depend on the thickness of the membrane. Then, the wafer was added into the NaOH solution that erodes SiO_2 on silicon, separating silicon and membrane with marked flakes. The membrane needs to be placed on the top of another wafer with the sample in the wet environment. When the water was evaporated, the membrane was fixed on top of the wafer. Corresponding etchants are used to melt the PMMA on the sample, and acetone is always used in the lab. The process is shown in Figure 7.4a. Although the liquid transfer method attains the stronger bonding between the layers, it is time-consuming and fails easily. Anhydrous method is the trend of 2D materials transfer process.

b) *Dry method*: Dry transfer method provides many benefits than wet transfer method. Actually, their principles are uniform but the environments they use are different. PMMA are used as a viscoelastic stamp in the wet method. In the dry method, samples stick on the stamp when vigorously pressed on the viscoelastic stamp. Then, precision alignment instrument would be used to align viscoelastic stamp and sample under the optical microscope. Finally, viscoelastic stamp is uncovered tardily from the stamp to obtain the heterostructures. The process of dry method is shown as Figure 7.4b. Moreover, repeating this procedure would result in more complicated heterostructures.

In the lab, PDMS (polydimethylsiloxane) are always used as the viscoelastic stamps because of the nontoxicity, nonflammability, high viscosity and excellent tenacity properties. In the process of making PDMS viscoelastic stamps, coagulant, thinner PDMS occupies a certain proportion, which leads to different viscosity, thickness, and tenacity. Diverse specialties of PDMS would affect transferring. Either liquid method or dry method is only fit for the vertical heterostructures.

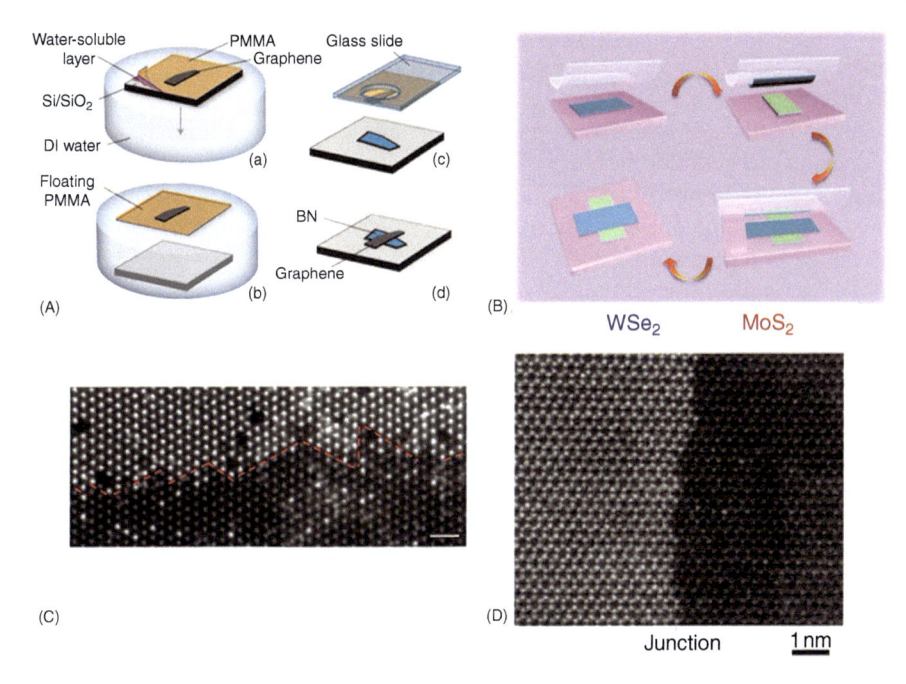

Figure 7.4 (A, B) Schematic diagrams of the steps of wet transfer and dry transfer, respectively. Source: (A) Reprinted with permission from Cui et al. [57]. Copyright 2017, Nature Publishing Group. (B) Reprinted with permission from Zhou et al. [58]. Copyright 2018, John Wiley & Sons. (C) and (D) Transmission electron microscopy (TEM) diagrams of one-step growth method and two-step growth method, respectively. Source: (C) Reprinted with permission from Gong et al. [46]. Copyright 2014, Nature Publishing Group. (D) Reprinted with permission from Li et al. [59]. Copyright 2014, Nature Publishing Group.

7.3.2 Chemical Methods

Physical transfer methods are convenient, fast, simple, and have repeatable properties; however, 2D heterostructures made by this approach may deform the samples and make them unstable. The specific lateral heterostructures require a spontaneous growth which the physical methods could not fabricate. The twist angle in the vertical heterostructures in the physical transfer methods is random and difficult to control. Therefore, the chemical methods are indispensible for the process of the fabrication of 2D heterostructures. There are several ways in chemical methods, including hydrothermal method and CVD.

The physical methods almost use the top-down strategy for the whole process, but chemical methods stacking layers use the down-top growing process in a spontaneous preferential growth. The contaminants would not induce the interface of different components easily in directly situ growth; however, chemical methods could facilitate intense coupling of the layers near the interface and obtain the lowest energy structures, which could improve the heterostructures and even their device characteristics.

There are two kinds of direct CVD growth methods as follows: one-step and two-step methods. In the one-step growth method, there is no such obvious boundary, and the whole process is a cogrowth pattern. However, in the two-step growth method, the two different components would be distinguished near the interface in a second. The two special CVD growth methods have their advantages and disadvantages. In the one-step growth method, contamination has no chance to introduction because of the whole growth cycle without the intermediate step in the process of 2D heterostructures fabrication. Meanwhile, many factors could not control artificially because of the intricate and heterogeneous elements in a cogrowth pattern. The successful 2D heterostructure preparation rather than the preparation of other components demands several essential factors such as the selection of precursors, growth temperature, growth time, growth procedure, the location of the powder and component design, etc. The 2D heterostructure fabrication by the one-step growth method is easier than that by the two-step growth method, according to the experiments. The two or more different components growing, respectively, would impede the compounding of alloy and atom interdiffusion between the components on the interface. The results of two kinds of methods are shown in Figure 7.4c,d. It is obvious that the interface is blurry through the one-step method and distinct through the two-step method. There is a huge disadvantage that contamination would mix in the growth procedure particularly between the steps that trigger to nucleation position in the interface changing the uniformity of 2D heterostructures [41]. Therefore, there are two ways to deplete the contamination induced. Firstly, the whole process of growth should not expose to the ambient atmosphere [42] or grow in the Ar/N_2 atmosphere. Otherwise, annealing is a neo-serviceable way to dislodge the contaminants after the steps [43].

The direct CVD growth paves the way to the large-area and industrialized fabrication of 2D heterostructures and is also a most probable path to realize. Nevertheless, in the process of direct CVD growth, many kinds of external factors influence the quality of 2D heterostructures easily. The direct CVD growth is still a challenge associated with extreme complexities of heterostructures. The typical 2D heterostructures could be classified into four types according to the material species: 2D semiconductor and 2D semiconductor, 2D semiconductor and h-BN, 2D semiconductor and graphene, h-BN, and graphene.

There are several factors to impact the process of 2D heterostructure growth and temperatures, substrates, precursors, and other factors, which would be discussed in this chapter.

a) *Temperatures*: High enough temperature is needed to vaporize solid precursors in the process of 2D heterostructure CVD growth. Moreover, different temperatures would impact the pressure of vapor precursors, thermodynamics, and kinetics to deposit the substrate [44, 45]. The deposition of rate and morphology could be controlled by different temperatures, and the vertical and lateral structures could also be controlled by temperatures. Gong et al. committed to the study of WS_2/MoS_2 heterostructures found that high temperatures about 850 °C have a tendency to the vertical structures in 2D heterostructure growth, and low temperatures about 650 °C tend to the lateral structures in 2D heterostructures [46]. Some characterization method can verify the morphology of the results. (i) Atomic force microscopy (AFM) could distinguish the sharp and boundaries to present a complete appearance. (ii) Raman spectroscopy is a good way to express the structural shape of materials because lateral or vertical structure could alter the optical regions. (iii) Scanning transmission electron microscopy (STEM) could further verify the structure of heterostructures standing by the sharp boundaries because 2H stacking of vertical structure and zigzag or armchair interface of lateral structures associated with a hexagonal lattice. The nucleation and growth rates of 2D compounds could distinguish the product of alloy, individual materials, or 2D heterostructures as shown in Figure 7.5a,b. Different elements would have different functions on the growing process actually and Figure 7.5c is as an example. Tellurium could help to melting the tungsten in the WS_2/MoS_2 heterostructures, which could facilitate the mix of the molybdenum trioxide (MoO_3), tungsten, and sulfur as precursors in one-step method. Gong et al. have explained the form of morphology in this heterostructure with changing temperatures [46]. MoS_2 possesses the easier nucleation and quicker growth than those of WS_2, therefore the priority growth of MoS_2 in the SiO_2/Si substrate as the vertical heterostructure. By contrast, when the edge of MoS_2 filled with dangling bonds easily bonded than the surface, the lower temperature impede of the nucleation and growth of WS_2 would be prone to combine with MoS_2, which almost leads to lateral structure. It can be seen that the vertical structure achieved by the higher temperature is more stable than the lateral structure formed by the low temperature. Because of the low-temperature kinetic effect and the high-temperature thermodynamic effect, the 2D heterostructure growth finds a suitable temperature point in 750 °C during the growth process.

In the two-step growth strategy, the effect of temperature on heterostructure growth is also very similar (shown as in Figure 7.5d). In addition, the concept of free energy change ($\Delta G_{r,\gamma}$) was described, as shown in the function (7.1), where the nucleation free energy change ($\Delta G_{r,\gamma}$) is related to the assumed nuclear radius (r) and surface energy composed of the surface energy of nucleus (γ_c), interfacial energy between the substrate and nucleus (γ_{sc}), surface energy of substrate (γ_s), and surface energy of monolayered nucleus edge ($\gamma_{c,\text{edge}}$).

$$\Delta G_{r,\gamma} = \pi r^2 t \Delta G_v + \pi r^2 (\gamma_c + \gamma_{sc} - \gamma_s) + 2\pi r t \gamma_{c,\text{edge}} \quad (7.1)$$

Here, t represents the thickness of the nucleus and ΔG_v means the difference in free energy per unit volume during phase change. It shows that the curve of the maximum free energy change (ΔG^*) depends on the radius of the nucleus (r). The monolayer atomic crystal and the nucleation curve on the SiO_2 substrate are shown in Figure 7.5e. Now that ΔG_v is replaced by $L\Delta T/T_m$, here, ΔT replaces $T_m - T_{\text{growth}}$. T_m and T_{growth} are the atomic crystal melt temperature and growth temperature, respectively. L expresses the latent heat for the reaction, and it can be seen that the degree of oversaturation of ΔT affects the positive correlation between ΔT and ΔG^*. Figure 7.5f shows that the nucleation rate is related to ΔT, which is approximately equal to the product of the positive correlation between the nuclear region and ΔT and the negative correlation between the thermal motion of the atom and ΔT, which determines

134 | *7 Properties of 2D Heterostructures*

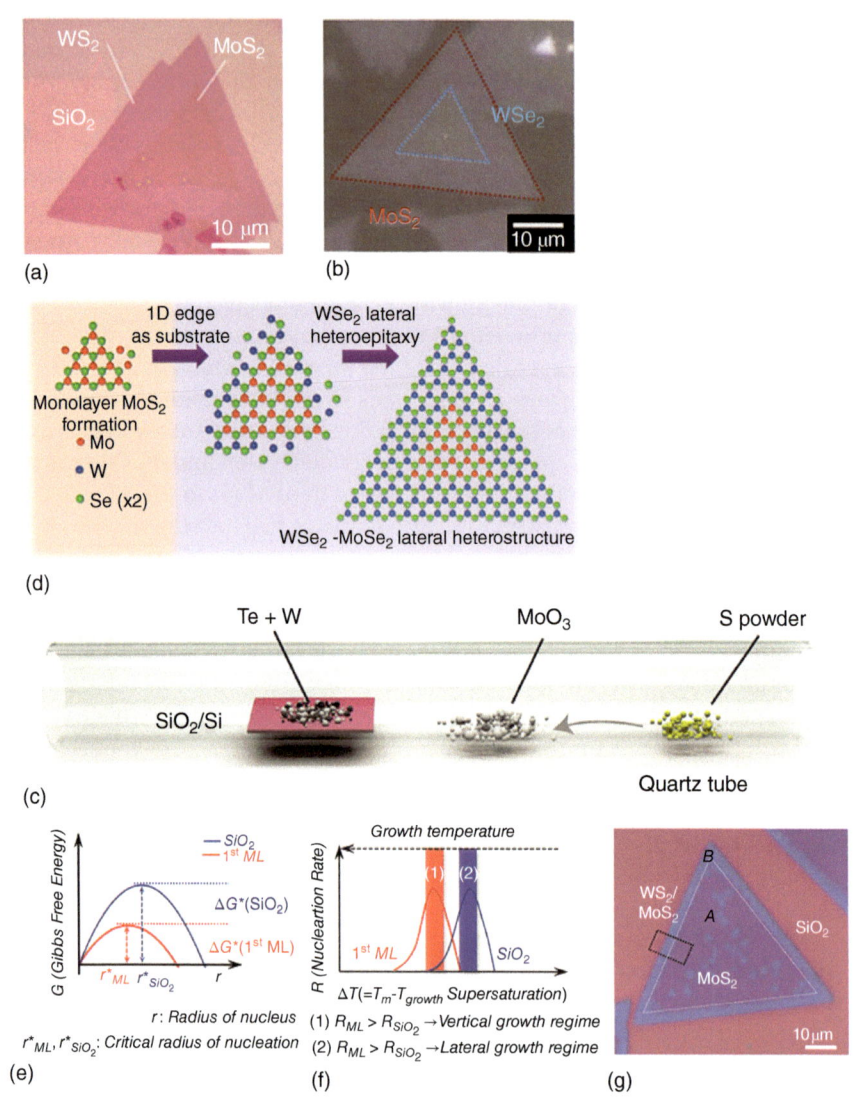

Figure 7.5 (a, b) Pictures of heterojunction under an optical microscope, Source: (a) Reprinted with permission from Gong et al. [46]. Copyright 2014, Nature Publishing Group. (b) Reprinted with permission from Li et al. [59]. Copyright 2014, Nature Publishing Group, (c) and the process of fabricating the WS_2/MoS_2 heterojunction via CVD. Source: Reprinted with permission from Zhou et al. [58]. Copyright 2018, John Wiley & Sons. (d) Growth process of two-dimensional $MoSe_2/WSe_2$ in-plane horizontal heterojunction. Source: Reprinted with permission from Huang et al. [52]. Copyright 2014, Nature Publishing Group. (e) The dependence curve of the Gibbs free energy curve on the nuclear radius. (f) The first-layer grows the 2D layered material in relation to the nucleation rate of the SiO_2 substrate surface. (g) An image of a vertical WS_2/MoO_2 heterojunction. Source: Reprinted with permission from Heo et al. [60]. Copyright 2018, John Wiley & Sons.

whether a vertical or lateral heterostructure is formed. Experiments show that temperature changes the growth morphology of the material through molecular dynamics. The result of 2D heterostructures is presented in Figure 7.5g. The difference between Figure 7.5a,g is as follows: the center of the MoS_2 monolayer area is diffused with small dots. The WS_2 in second growth covers several microns of the triangular monolayer MoS_2 grown in the first growth. Figure 7.5g gives an explanation that these dots are the first-growth bilayered composition with the lateral epitaxial second-growth monolayered composition. According to the experimental verification, in the one-step growth, the first-layer composition may preferentially occupy and simultaneously grow on the edge of the first layer because of the dangling bonds at the edges; therefore, the two-step method forces the second-layer composition to grow in the center of the layer surface and caused the extension. In addition, the temperature has a considerable influence on the growth process of the 2D heterostructures. The temperature of precursor vapor controls the thickness of MoS_2 stacks on other semiconductors (such as SnS_2, TaS_2, and graphene) and associated with the reaction time [47]. In summary, temperature as a controllable variable determines the structure and morphology of 2D heterostructure growth.

b) *Substrates*: The Si/SiO_2 substrate is the best substrate choice for compatibility with contemporary semiconductor technology. Therefore, when selecting a substrate for 2D heterostructure growth considering the combination with the current integrated circuit process, the Si/SiO_2 substrate is preferred. However, when MoS_2 was successfully grown on SiO_2/Si substrates under normal CVD growth conditions, a series of problems began to attract people's attention [48]. Some of the more excellent substrates are emerging, with better flexibility and controllability during CVD growth. For example, annealing sapphire substrates more facilitated for the growth of uniform lattice orientation of MoS_2 [49], and large-size monolayered $MoSe_2$ is easier to grow on the glass substrates [50]. This section will focus on how to select a suitable substrate in a 2D heterostructures in the CVD growth process.

The quality of materials grown on different substrates is discrepant. There is an example for the heterostructures grown on the different substrates-SiO_2/Si and h-BN substrates. Heo et al. investigated that Bi_2Te_3 powder and Sb_2Te_3 powder as precursors, respectively, to grow two different crystalline morphologies of Bi_2Te_3/Sb_2Te_3 heterostructures on SiO_2/Si and h-BN substrates by two-step low-pressure physical vapor deposition process. Bi_2Te_3 and Sb_2Te_3 grown on SiO_2/Si substrate always present as triangular morphologies in a 3D island, as shown in Figure 7.6a,b. However, the Bi_2Te_3 and Sb_2Te_3 grown on h-BN substrate always present as irregular morphologies in a layer-by-layer as shown in Figure 7.6c,d [51].

The growth pattern of the heterostructures on the substrates is related to the adatom diffusion velocities. The roughness surface and dangling bonds limit the diffusion of precursors on SiO_2. Nevertheless, because of the relationship of $D_{s,\text{h-BN}} > D_{s,Bi_2Te_3} > D_{s,SiO_2}$, the triangular morphology of the Bi_2Te_3/Sb_2Te_3 heterostructures on the SiO_2/Si substrate is grown, and the amorphous crystalline morphology is grown on the h-BN substrate. Conversely, the precursor is more readily deposited on the h-BN substrate than the surface

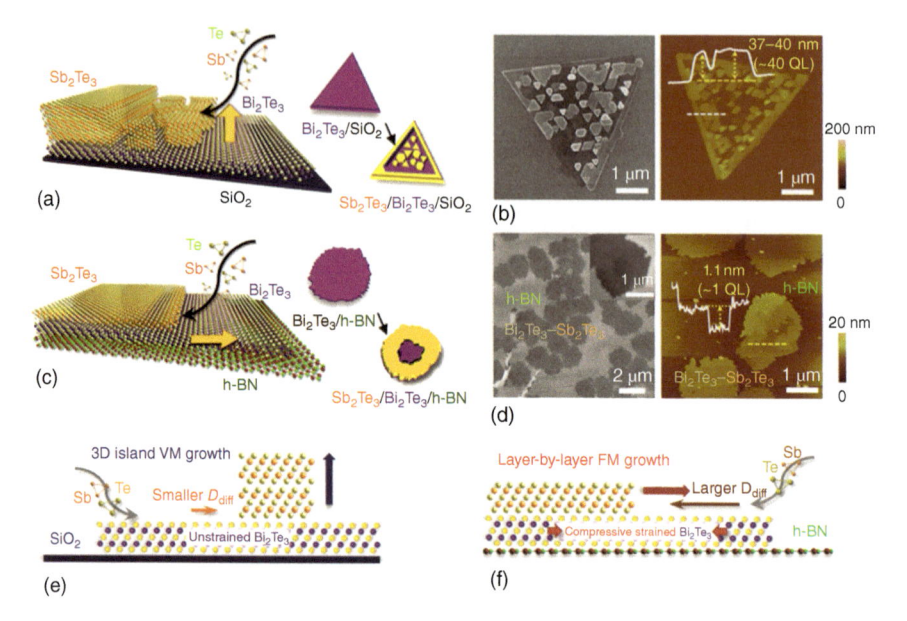

Figure 7.6 (a, b) Growth process, optical image, and AFM image of growing vertical heterojunction on SiO_2 substrates. (c, d) Growth process, optical image, and AFM image of growing vertical heterojunction on h-BN substrates. (e, f) Growth of the second-layer Sb_2Te_3 on SiO_2 and h-BN substrates, respectively. Source: Cui et al. 2018 [57]. Reproduced with permission of Springer Nature.

of the material at a faster rate of adsorption of atoms, thereby forming amorphous crystal morphology. Because of the residual stress, the crystal morphology of the upper layer changes on the heterostructure surface. As shown in Figure 7.6e, Sb_2Te_3 on unstrained Bi_2Te_3 was assembled in a 3D island growth mode, and the compressive strain Bi_2Te_3 with enhanced adsorption atomic diffusion as shown in Figure 7.6f exhibited layer-by-layer growth. Sometimes, a suitable material substrate should be chosen to reduce the barrier energy of heterostructure growth.

c) *Precursors*: Precursors play an important role in the process of the fabrication of 2D heterostructures. Different precursors need to control different conditions, and the quality of the heterostructures is different. Involuntary, the same precursor will also get different results according to different conditions. Even precursors of the same species and different stoichiometry will get different material quality. Specifically in the fabrication of lateral heterostructures, the accurate stoichiometry precursors would need to control the quality of bottom 2D material in the first-growth step so that the second-growth step could deposit along the edge of the first-growth portion well. Huang et al. investigated the effect of stoichiometric ratio of precursor on $MoSe_2/WSe_2$ heterojunction via one-step low-pressure physical vapor deposition [52]. The results indicate that only WSe_2 was obtained when $WSe_2 : MoSe_2 = 0.06\,g : 0.02\,g$. As the composition of WSe_2 gradually decreases ($0.03\,g : 0.05\,g$), the range of WSe_2 outer in the heterojunction also gradually decreases. When the ratio is $WSe_2 : MoSe_2 = 0.06\,g : 0.05\,g$, the heterostructure of $MoSe_2/WSe_2$ has obtained uniform appearance. In the growing process, $MoSe_2$ is first deposited as a triangle on the silicon wafer because $MoSe_2$ deposits faster than WSe_2. Until a certain ratio, $MoSe_2$ stops growing than WSe_2 deposit. Therefore, the stoichiometric ratio of precursors would impact the quality of 2D heterostructures extremely.

In addition, the control of the precursor proportion can also result in the formation of 2D heterostructures or an alloy [53]. By changing the composition of the precursor, it is sometimes possible to inhibit the formation of the alloy, thereby forming a 2D heterostructure. The addition of special W–Te precursors to WS_2 and MoS_2 in the production of WS_2/MoS_2 heterostructures can always restrain the formation of alloys, which was investigated in Ref. [46]. The type and quantity of precursors are easily and precisely controlled, so controlling the precursors can achieve the purpose of optimizing 2D heterostructure formation. For instance, Zhang et al. investigated that to compare three kinds of materials including MoX_2 (X = S, Se), MoO_3, and thermal evaporated MoO_3-x thin layers on Si for the lateral WX_2/MoX_2 heterostructures, only MoX_2 precursors and thermal evaporated MoO_3-x thin layers on Si could fabricate MoX_2. Moreover, precursors in different vapor pressures still control the growth of heterojunctions [54].

It can be found that the types of precursors are basically not limited and can be classified into pure metals, metal oxides, sulfides, etc., by material type and can be classified into nanowires, flakes, powders, etc., in terms of morphology of the materials. In addition, there are two kinds of methods based on film growth. Firstly, the film precursor, like the conventional precursor, participates in the synthesis as part of the heterostructures. The other is that the precursor is directly sulfurized to facilitate a heterostructure. However, this method is far less crystalline than the growth of heterostructures by CVD.

d) *Other factors*: During the growth of the heterostructures, there are other factors that affect the growth of the heterostructures. First, the problem of lattice mismatch has always hindered the development of heterostructures. In vertical heterostructures, the layers are connected by van der Waals forces, and lattice mismatch is less of a concern. In horizontal heterostructures, different materials are connected laterally by covalent bonds, so the lattice mismatch has a greater impact on horizontal heterostructures. Xi Ling et al. compared the growth quality of the three lateral heterojunctions of graphene–MoS_2, WS_2–MoS_2, and h-BN–MoS_2, respectively [55]. After comparison, graphene–MoS_2 heterojunction was selected as the base for the device, in which the stress directly affected the quality of the heterojunction. The result could be found in Figure 7.7.

7.4 Characterizations of 2D Heterostructures

2D heterostructures combined the advantages of several 2D materials and are even superior to the one-component materials because of the band offset and interactions among the components and possess extensive applications

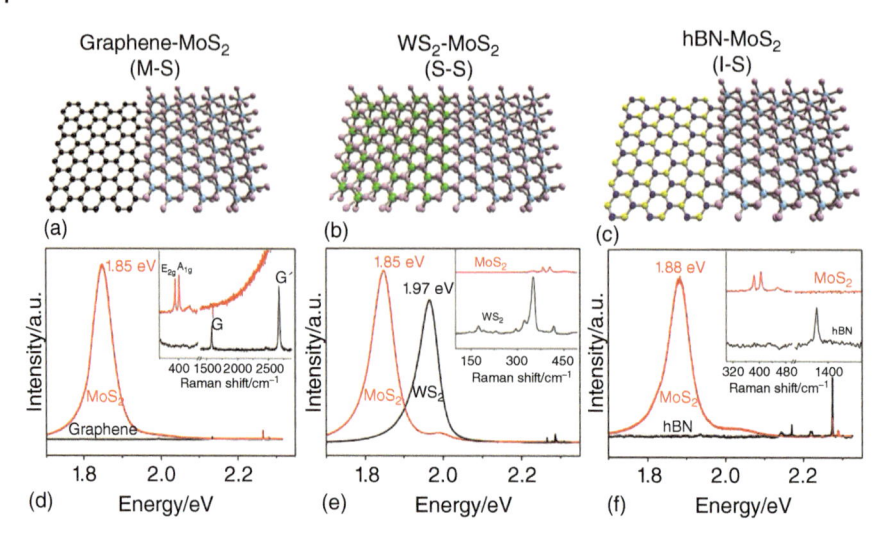

Figure 7.7 (a–c) Schematic illustration of graphene–MoS$_2$, WS$_2$–MoS$_2$, and h-BN–MoS$_2$ lateral heterojunction, respectively. (d–f) PL spectrum and Raman spectroscopy of graphene–MoS$_2$, WS$_2$–MoS$_2$, and h-BN–MoS$_2$ lateral heterojunction. Source: Reprinted with permission from Ling et al. [55]. Copyright 2016, John Wiley & Sons.

in special circumstance. The bending of band in diverse 2D heterostructures deeply influences the properties of electrical, optical, and photoelectric. 2D heterostructures could be classified into three types on the basis of conduction band and valence band matching in the different 2D components. Similar to traditional heterostructures, there are type I, II, and III, as shown in Figure 7.8.

Type I: The conduction band and valence band in the narrow bandgap 2D materials are within the conduction band and valence band in the wide bandgap components.

Type II: Two kinds of type II 2D heterostructures would be discussed. In type IIA, the conduction band in one component lower than conduction band in another component, and also the valence band in the component is lower than the valence band in another component. The bandgap in type IIA is stagger. However, the bandgap staggering is bigger between the two components. The bottom of conduction band and the top of valence band in the narrow bandgap component are within the valence band in the wide bandgap component. The band alignment is beneficial for ultrafast separation of carriers and attains high photoelectric conversion efficiency.

Type III: One of the 2D materials is zero bandgap, such as grapheme. Grapheme possesses excellent conductivity, thus playing a good role of electrode in majority situations.

Interlayer coupling is strongly influenced by the approach of transfer. For example, strong interlayer coupling because of spontaneous stacking configurations is remarkable by direct CVD method. Raman spectroscopy and photoluminescence spectroscopy are used to represent the interlayer coupling of 2D heterostructures generally. For example, the band alignment, phonon modes,

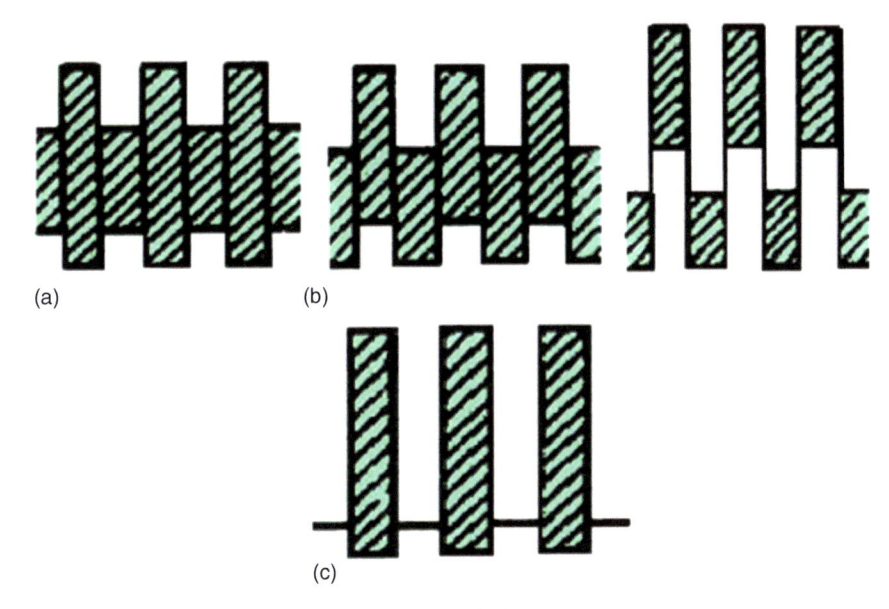

Figure 7.8 (a) Type I, (b) type IIA and type IIB, and (c) type III.

Raman spectra, photoluminescence spectra, and photoluminescence intensity mapping of coupled always serve as the methods for investigating the electrical and optical properties of 2D materials. Interlayer coupling would cause the shift of phonon vibration modes and distinct photoluminescence (PL) peaks.

Experiments indicated that the light absorption of 2D heterostructures is higher than that of one-component materials and have extensive applications in photoelectric detectors, photovoltaic devices, and electric devices. 2D heterostructures change the band alignment of original component associated with the band mismatching that the lattice constant of different 2D materials are disparate.

Selecting the appropriate 2D materials with inverse types of carriers could compose p–n junctions that exhibit the favorable rectification property. In addition, Esaki diodes was indicated the negative differential resistance in plentiful works. The contact of 2D components with graphene facilitates the flowing of carriers than that with metal attributing to the matching of band alignment; therefore, the heterostructures could replace the metal electrode.

References

1 Novoselov, K.S., Geim, A.K., Morozov, S.V. et al. (2004). *Science* 306: 666.
2 Castro Neto, A.H., Guinea, F., Peres, N.M.R. et al. (2009). *Rev. Mod. Phys.* 81: 109.
3 Schwierz, F. (2010). *Nat. Nanotechnol.* 5: 487.
4 Bonaccorso, F., Sun, Z., Hasan, T., and Ferrari, A.C. (2010). *Nat. Photonics* 4: 611.

5 Balandin, A.A., Ghosh, S., Bao, W. et al. (2008). *Nano Lett.* 8: 902.
6 Lee, C., Wei, X.D., Kysar, J.W., and Hone, J. (2008). *Science* 321: 385.
7 Zhang, H., Liu, C.-X., Qi, X.-L. et al. (2009). *Nat. Phys.* 5: 438.
8 Lima, E.N., Schmidt, T.M., and Nunes, R.W. (2016). *Nano Lett.* 16: 4025.
9 Tsen, A.W., Hunt, B., Kim, Y.D. et al. (2015). *Nat. Phys.* 12: 208.
10 Molle, A., Goldberger, J., Houssa, M. et al. (2017). *Nat. Mater.* 16: 163.
11 Mannix, A.J., Kiraly, B., Hersam, M.C., and Guisinger, N.P. (2017). *Nat. Rev. Chem.* 1: 0014.
12 Li, S.L., Tsukagoshi, K., Orgiu, E., and Samori, P. (2016). *Chem. Soc. Rev.* 45: 118.
13 Shi, J., Tong, R., Zhou, X. et al. (2016). *Adv. Mater.* 28: 10664.
14 Shi, J., Ji, Q., Liu, Z., and Zhang, Y. (2016). *Adv. Energy Mater.* 6: 1600459.
15 Chhowalla, M., Shin, H.S., Eda, G. et al. (2013). *Nat. Chem.* 5: 263.
16 Li, L.H. and Chen, Y. (2016). *Adv. Funct. Mater.* 26: 2594.
17 Prete, M.S., Mosca Conte, A., Gori, P. et al. (2017). *Appl. Phys. Lett.* 110: 012103.
18 Osada, M. and Sasaki, T. (2009). *J. Mater. Chem.* 19: 2503.
19 Ma, R.Z. and Sasaki, T. (2015). *Acc. Chem. Res.* 48: 136.
20 Anasori, B., Lukatskaya, M.R., and Gogotsi, Y. (2017). *Nat. Rev. Mater.* 2: 16098.
21 Naguib, M., Mochalin, V.N., Barsoum, M., and Gogotsi, Y. (2014). *Adv. Mater.* 26: 982.
22 Liu, H., Neal, A.T., Zhu, Z. et al. (2014). *ACS Nano* 8: 4033.
23 Tai, G.A., Hu, T.S., Zhou, Y.G. et al. (2015). *Angew. Chem. Int. Ed.* 54: 15473.
24 Fleurence, A., Friedlein, R., Ozaki, T. et al. (2012). *Phys. Rev. Lett.*: 108.
25 Li, L.F., Lu, S.Z., Pan, J.B. et al. (2014). *Adv. Mater.* 26: 4820.
26 Zhu, F.F., Chen, W.J., Xu, Y. et al. (2015). *Nat. Mater.* 14: 1020.
27 Zhang, S.L., Yan, Z., Li, Y.F. et al. (2015). *Angew. Chem. Int. Ed.* 54: 3112.
28 Ji, J.P., Song, X.F., Liu, J.Z. et al. (2016). *Nat. Commun.* 7: 13352.
29 Zhao, C., Xu, Z., Wang, H. et al. (2014). *Adv. Funct. Mater.* 24: 5985.
30 Tsipas, P., Kassavetis, S., Tsoutsou, D. et al. (2013). *Appl. Phys. Lett.* 103: 251605.
31 Ong, W.-J., Tan, L.-L., Ng, Y.H. et al. (2016). *Chem. Rev.* 116: 7159.
32 Al Balushi, Z.Y., Wang, K., Ghosh, R.K. et al. (2016). *Nat. Mater.* 15: 1166.
33 Lu, A.Y., Zhu, H.Y., Xiao, J. et al. (2017). *Nat. Nanotechnol.* 12: 744.
34 Roy, T., Tosun, M., Cao, X. et al. (2015). *ACS Nano.* 9: 2071.
35 Sarkar, D., Xie, X.J., Liu, W. et al. (2015). *Nature* 526: 91.
36 Deng, Y.X., Luo, Z., Conrad, N.J. et al. (2014). *ACS Nano.* 8: 8292.
37 Xue, Y.Z., Zhang, Y.P., Liu, Y. et al. (2016). *ACS Nano.* 10: 573.
38 Ye, L., Wang, P., Jin Luo, W. et al. (2017). *Nano Energy*: 37, 8292–8299.
39 Cheng, R., Li, D.H., Zhou, H.L. et al. (2014). *Nano Lett.* 14: 5590.
40 Withers, F., Del Pozo-Zamudio, O., Mishchenko, A. et al. (2015). *Nat. Mater.* 14: 301.
41 Li, Y.B., Zhang, J.S., Zheng, G.Y. et al. (2015). *ACS Nano.* 9: 10916.
42 Duan, X.D., Wang, C., Shaw, J.C. et al. (2014). *Nat. Nanotechnol.* 9: 1024.
43 Gong, Y.J., Lei, S.D., Ye, G.L. et al. (2015). *Nano Lett.* 15: 6135.

References

44 Gulbransen, E.A., Andrew, K.F., and Brassart, F.A. (1963). *J. Electrochem. Soc.* 110: 952.
45 Tromp, R.M. and Hannon, J.B. (2009). *Phys. Rev. Lett.*: 102, 106104.
46 Gong, Y.J., Lin, J.H., Wang, X.L. et al. (2014). *Nat. Mater.* 13: 1135.
47 Samad, L., Bladow, S.M., Ding, Q. et al. (2016). *ACS Nano.* 10: 7039.
48 Zhang, J., Wang, J.H., Chen, P. et al. (2016). *Adv. Mater.* 28: 1950.
49 Dumcenco, D., Ovchinnikov, D., Marinov, K. et al. (2015). *ACS Nano* 9: 4611.
50 Chen, J.Y., Zhao, X.X., Tan, S.J.R. et al. (2017). *J. Am. Chem. Soc.* 139: 1073.
51 Heo, H., Sung, J.H., Ahn, J.-H. et al. (2017). *Adv. Elect. Mater.* 3: 1600375.
52 Huang, C., Wu, S., Sanchez, A.M. et al. (2014). *Nat. Mater.* 13: 1096.
53 Obraztsov, A.N. (2009). *Nat. Nanotechnol.* 4: 212.
54 Zhang, X.Q., Lin, C.H., Tseng, Y.W. et al. (2015). *Nano Lett.* 15: 410.
55 Ling, X., Lin, Y.X., Ma, Q. et al. (2016). *Adv. Mater.* 28: 2322.
56 Duesberg, G.S. (2014). *Nat. Mater.* 13: 1075.
57 Cui, Y., Li, B., Li, J.B., and Wei, Z.M. (2018). *Sci. China Phys. Mech. Astron.* 61: 016801.
58 Zhou, X., Hu, X., Yu, J. et al. (2018). *Adv. Funct. Mater.* 28: 1706587.
59 Li, M.Y., Shi, Y.M., Cheng, C.C. et al. (2015). *Science* 349: 524.
60 Heo, H., Sung, J.H., Jin, G. et al. (2015). *Adv. Mater.* 27: 3803.

8
Application in (Opto) Electronics

8.1 Field-Effect Transistors

An field-effect transistor (FET) is a kind of device in which an electric field can capacitively control channels. According to the type of channel carrier, there are n-channel devices and p-channel devices [1]. The carrier in the channel of an n-channel device is the electron. Positive gate voltage results in enhanced conductivity. The carrier in the channel of a p-channel device is the hole, and more negative gate voltage leads to stronger conductivity. FETs are widely used as basic units of modern integrated circuits by virtue of their simple structure and low energy consumption. In recent years, the development directions of FETs, like other electronic components, are miniaturization and high performance [2]. However, in order to achieve miniaturizing FETs, when the size reaches a few nanometers, the traditional silicon-based channel materials are obviously limited by the short channel effect [3]. Because of their unique atomic structure, two-dimensional materials are a promising option to overcome the short channel effect and effectively control electrons, so the application of two-dimensional materials in FETs has a broad prospect. Graphene has been widely concerned because of their ultrahigh electron and hole mobility, but their performance is greatly limited by the low switch ratio caused by the gapless band structure of graphene. In 2014, L. Li et al. reported the first few-layer black phosphorus (bP)-based FETs, as shown in Figure 8.1a. From the transfer curves (Figure 8.1b), the switching ratio can reach the order of 10^5, and the charge carrier mobility is found to be thickness-dependent, with the highest values up to $\sim 1000\,cm^2/V\,s$ obtained for a thickness of $\sim 10\,nm$, which demonstrates the potential of black phosphorus thin crystals as a new two-dimensional material for applications in nanoelectronic devices [4].

On the other hand, the main problem of bP-based FETs is the instability of black phosphorus in air. Therefore, many 2D materials with the natural bandgap and good stability have been gradually applied to FETs. 2D transition metal dichalcogenides (TMDs) are representative. Although the unique atomic structure of two-dimensional metal chalcogenides poses some challenges to the fabrication and processing of devices, this does not prevent FET channels based on 2D TMDs from being widely studied and applied. So far, many two-dimensional metal chalcogenides have been applied to FETs, such as typical

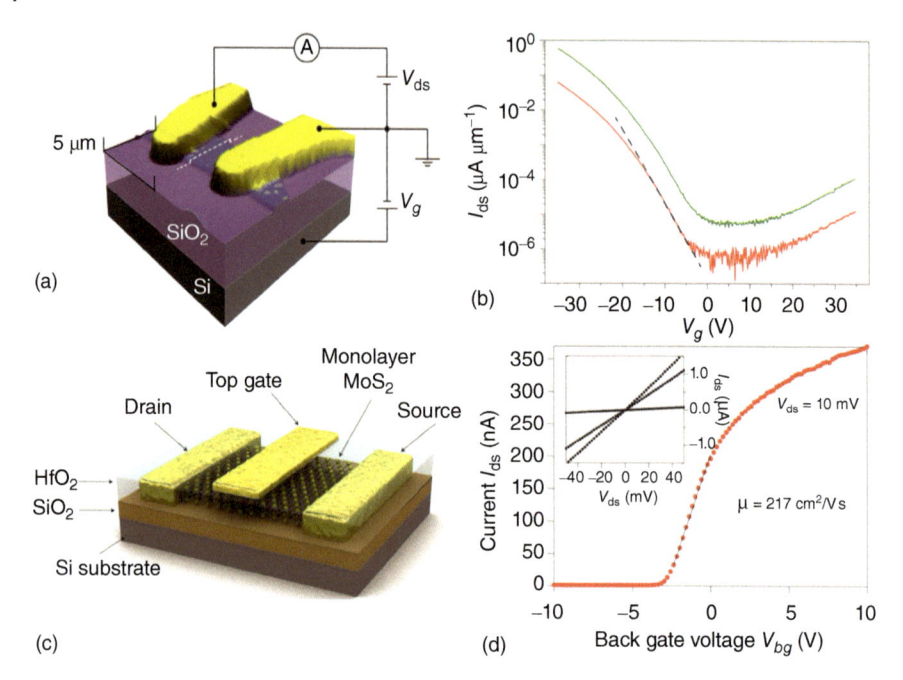

Figure 8.1 (a) Schematics of bP-based transistors. (b) Transfer curves of bP transistors. Source: Reprinted with permission from Li et al. [4]. Copyright 2014, Springer Nature. (c) Schematics of monolayer MoS_2-based transistors. (d) Transfer curves of the MoS_2 transistors. Source: Reprinted with permission from Radisavljevic et al. [5]. Copyright 2011, Springer Nature.

MoS_2 [5], WS_2 [6], InSe [7, 8], and so on. MoS_2 was firstly applied to the channel material because its single-layer material has a bandgap of 1.8 eV [5, 9]. In 2011, A. Kis and co-workers group developed the first MoS_2 FETs [5], and the schematics of single-layer MoS_2 FETs are shown in Figure 8.1c. They used a hafnium oxide gate dielectric to demonstrate a room temperature single-layer MoS_2 mobility of at least 200 cm^2/V s, with room temperature current on/off ratios of 10^8 (Figure 8.1d). Monolayer MoS_2 has a direct bandgap, which can offer lower power consumption than classical transistors. Compared with the low mobility of other two-dimensional metal chalcogenides, InSe showed a high mobility. D. A. Bandurin et al. reported a high-quality 2D electron gas in few-layer InSe encapsulated in hexagonal boron nitride under an inert atmosphere. Carrier mobilities are found to exceed 10^3 and 10^4 cm^2/V s at room and liquid helium temperatures, respectively, allowing the observation of the fully developed quantum Hall effect [8]. Its high mobility was comparable to silicon film and black phosphorus film, and its stability was better than that of black phosphorus [8]. In addition, many two-dimensional metal chalcogenides with anisotropic crystal structures exhibit anisotropic electrical properties, such as ReS_2 and $ReSe_2$ [10].

In the design of FETs, it is feasible to select two-dimensional metal chalcogenides as channel materials, which can meet the basic requirements of the current switching. Two-dimensional metal chalcogenides generally have the high theoretical carrier mobility, such as the theoretical carrier mobility of MoS$_2$ can reach 410 cm^2/V s [11]. However, as a result of the two-dimensional metal chalcogenide, the carrier mobility is easily affected by electrode contact and multiple scattering mechanisms (such as internal defect, external Coulomb scattering, and interface phonon scattering), so usually single-layer MoS$_2$-based FETs cannot reach as high as its theoretical value of carrier mobility. However, many attempts have been made to improve the carrier mobility of FETs based on these materials to reach the inherent carrier mobility level of materials. For example, high dielectric constant materials such as HfO$_2$ and Al$_2$O$_3$ are used as the dielectric layer of FETs. In this way, the influence of Coulomb scattering on carrier mobility can be shielded as far as possible, so as to effectively improve the carrier mobility of FETs. The phonon scattering of dielectric layer should be considered in addition to Coulomb scattering. Because of its good compatibility and its ability to shield phonon scattering and Coulomb scattering effectively, h-BN has become a good insulation material for two-dimensional metal chalcogenides. In 2018, N. Huo et al. reported the chemical vapor deposition (CVD) grown monolayer MoS$_2$ FETs with large mobility improvement by employing a facile method to engineer dielectric environment and phonon suppression via atomic layer deposition (ALD) technique. They found that both Coulomb scattering and phonon scattering can be suppressed upon Hafnia covering, leading to the much improved mobility [12].

In the practical application of FETs, researchers have made some attempts, such as ultrashort channel devices, logic circuits, memories, and so on. The gate length is an important parameter in modern integrated circuits. It is usually only a few nanometers. The carriers in two-dimensional metal chalcogenides are distributed in atomic-scale thickness, which means that they have advantages on the gate control compared with conventional bulk materials. A two-dimensional metal chalcogenide FET also provides a more flexible strategy for analog circuits. For example, referring to nonvolatile memory applications, the memory cell with the traditional flash memory structure based on monolayer MoS$_2$ and few-layered graphene as the channel material and floating gate exhibit strong charge storage ability.

8.2 Infrared Photodetectors

Photodetectors that convert light into electrical signals are one of the key components in modern multifunctional technologies. Traditional crystalline silicon photodetectors have reached a high level of maturity and performance in digital cameras and optical sensing systems. Infrared photodetectors based on epitaxially grown InGaAs, InAs, HgCdTe, and type-II superlattices have also allowed sensing in the infrared spectrum beyond silicon's reach. However, silicon, having an indirect bandgap, is a poor light absorber, and thick silicon photodetectors are characterized by their bulky, rigid, and brittle form factor, preventing their specific applications from flexible optoelectronic platforms. In the infrared photodetectors, InGaAs and other exotic semiconductors possess similar drawbacks in addition to the complementary metal oxide semiconductor (CMOS) incompatibility and significant cost because of the complex manufacturing. Two-dimensional (2D) atomic sheets with intralayer covalent bonding and interlayer van der Waals (vdW) interaction have emerged as a unique and promising material family for photonics and optoelectronics in view of their appealing characteristics, compared to conventional materials. In this section, we will discuss the figure of merit, photodetection mechanism, progress, and outlook of 2D-based infrared photodetectors.

8.2.1 Figures of Merit

Photon flux (ϕ_p) is the number of incident photons per second on a detector, which is the incident optical power divided by a phonon energy $\phi_p = \frac{P_{in}}{h\nu}$, where P_{in} is the incident optical power, h is Planck's constant, and ν is the frequency of incident photon.

Electron flux (ϕ_e) is the number of electron–hole pairs per second collected to produce photocurrent $\phi_e = \frac{I_{ph}}{e}$, where I_{ph} is the photocurrent and e is the electron charge.

Responsivity (R) is defined as $R = \frac{I_{ph}}{P_{in}}$, describing a photocurrent flowing in a detector divided by incident optical power in the units of A/W.

External quantum efficiency (EQE), is the ratio of the electron flux in a detector to the total incident phonon flux, $EQE = \frac{\phi_e}{\phi_p} = \frac{I_{ph}}{P_{in}} \frac{h\nu}{e} = R\frac{h\nu}{e}$

Internal quantum efficiency (IQE), is the ratio of the electron flux to the absorbed phonon flux, $IQE = \frac{\phi_e}{A_a \phi_p} = R\frac{h\nu}{A_a e}$, where A_a is the absorbed fraction.

Photo-gain (G), describes the number of photoexcited electron–hole pairs per single incident photon. In a photodiode, the photogain is equal to unity, unless carrier multiplication effects are present. In a photoconductor, one type of carrier (say hole) is usually captured in trap states or sensitizing centers with a lifetime of τ_{life}, while the other type of carrier (say electron) is free to traverse the channel with a transit time of $\tau_{transit}$. If carrier lifetime is longer than transit time, the free electrons recirculate many times before recombination with captured holes, leading to a generation of photogain, which is defined by $G = \frac{\tau_{life}}{\tau_{transit}} = \frac{\tau_{life}}{L^2} \mu V_{DS}$, where L is length of the channel, μ is the carrier mobility, and V_{DS} is the applied bias across channel.

Time constant (τ): Time constant represents the time the detector needs to reach $1/e$ (~37%) of its initial value and can be identified by fitting the current response using exponential function: $I(t) = I_0 \left(1 - e^{-\frac{t}{\tau}}\right)$ for current rise or $I(t) = I_0 e^{-\frac{t}{\tau}}$ for current decay, where I_0 is the initial value (here is electrical current under light illumination).

Rise time (t_r) and decay time (t_f): Rise time and decay time are also usually used to characterize the response speed, and they are defined as the time interval required for the response to rise (decay) from 10% (90%) to 90% (10%) of its initial value. The relationship between time constant and rise time (or decay time) is t_r (or t_f) $\approx 2.2\tau$.

Bandwidth (*B*) in hertz (Hz) is the frequency range of modulated light that a detector can fully follow. It is determined by the time constant: $B = \frac{1}{2\pi\tau}$.

Noise current (*I*_{noise}), is the random root mean square fluctuation in current when bandwidth is limited to 1 Hz. The unit is A Hz$^{-1/2}$. The magnitude of noise current can be measured via a lock-in amplifier or extracted from the noise spectral density S_n, which is obtained by calculating the Fourier transformation of dark current traces.

Signal-to-noise ratio (*SNR*) is the ratio of signal power-to-noise power, which must be larger than unity so that the signal power can be distinguished from the noise.

Noise equivalent power (*NEP*), is the minimum detectable optical power in watts at which the electrical *SNR* in the detector is equal to unity, when bandwidth is limited to 1 Hz. The unit is W Hz$^{-1/2}$. *NEP* describes the sensitivity of a detector and is determined by both noise current and responsivity, $NEP = \frac{I_{noise}}{R}$.

Detectivity (*D**) is another common figure of merit to characterize the sensitivity of a detector, which can enable the comparison of detectors with different geometries. The unit is cm Hz$^{1/2}$/W or Jones. It can be defined by $D^* = \frac{\sqrt{AB}}{NEP} = \frac{R\sqrt{AB}}{I_{noise}} = \frac{R\sqrt{A}}{S_n}$, where *A* is active area, *B* is electrical bandwidth, and S_n is noise spectral density.

8.2.2 Photodetection Mechanism

8.2.2.1 Photothermoelectric Effect

Photothermoelectric effect is generated by the temperature gradient ΔT between different substances. A photovoltage V_{PTE} can be produced through the Seebeck effect: $V_{PTE} = (S_1 - S_2)\Delta T$, where S_1 and S_2 (in V/K) are the Seebeck coefficients of the two substances (Figure 8.2a). The hot carriers can populate under the photovoltage to form current at zero external bias. The photothermoelectric effect (PTE) effect plays a key role in the photoresponse of graphene detectors because of the dominated hot carrier transport and strong e–e interactions in graphene [14]. The MoS$_2$ detectors were also reported to present a significant PTE effect arising from the large mismatch of Seebeck coefficients between MoS$_2$ and metal electrodes [15]. It is reported that the Seebeck coefficients of MoS$_2$ (30–100 mV/K) are several orders of magnitude larger than in graphene (4–100 μV/K).

8.2.2.2 Bolometric Effect

The underlying physical mechanism of the bolometer is bolometric effect, which is used to detect the changes in incident photon radiation (*dP*) by measuring the changes in temperature (*dT*), the bolometric effect is associated with the conductance change (*dG*) of the semiconductor channel induced by uniform light heating on homogeneous material (Figure 8.2b) [13]. The sensitivity of bolometer is ultimately determined by the thermal conductance: $G_h = dP/dT$. Upon light irradiation, the temperature changes as a function of time with thermal time constant $\tau = C_h/G_h$, where C_h is the heat capacity that determines the response time.

(a)   (b)

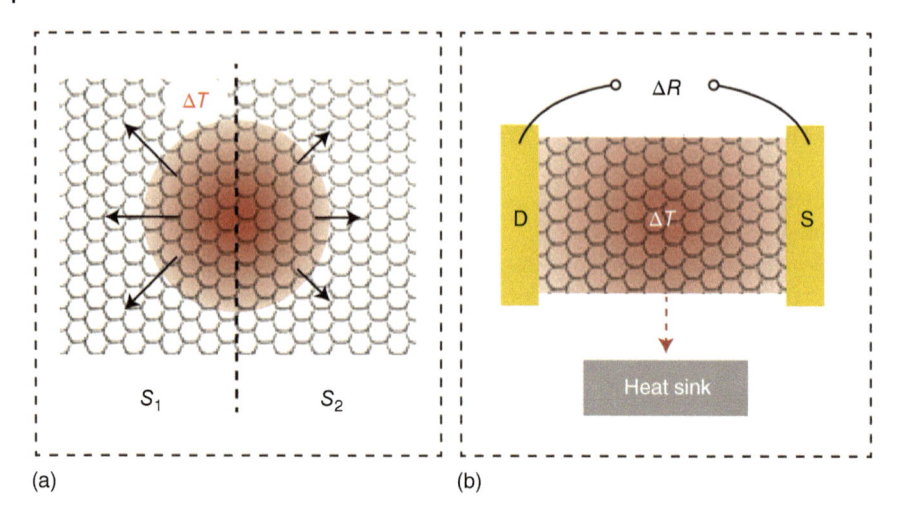

Figure 8.2 Schematic representation of the photothermoelectric (a) and bolometric (b) effect in graphene detectors. Source: Reprinted with permission from Koppens et al. [13]. Copyright 2014, Springer Nature.

8.2.2.3 Photogating Effect

In the photogating effect, the electron–hole pairs are generated under light illumination, and subsequently, one of the two carrier types (say electrons) is trapped in the localized states located at defects or sensitizers, while another type of carrier (say holes) is transferred to the channel. These trapped electrons act as local gates and effectively modulate the conductance of the 2D materials because of electrostatic interactions. The holes in the channel can recirculate many times during the lifetime of the trapped carriers, leading to a high gain. On the other hand, the long lifetime (i.e. high gain) is at the expense of lower temporal bandwidth typically below 1 kHz, which can be applied in detectors that require high sensitivity but not very fast operation such as video imaging.

8.2.2.4 Photovoltaic Effect

The photovoltaic effect relies on a built-in electric field to separate the photogenerated electron–hole pairs. The built-in electric field is normally produced at a junction, where there is a significant difference in the work functions between materials, such as p–n or Schottky junctions. Photodetectors operating via the photovoltaic effect are called photodiodes. In photovoltaic mode, the photogenerated electron–hole pairs are separated by the built-in electric field, and the electrons and holes are collected at opposite electrodes, forming a significant short-circuit current. In a photodiode, the dark current is very low at zero bias, leading to much low noise and high sensitivity. By applying large reverse bias, avalanche multiplication can occur, providing large internal gain.

8.2.2.5 Plasmonic Effect

In metal nanostructures, the coherent oscillations of free electrons existed as plasmons that can couple to the incident light field and provide intense electromagnetic field concentration from a wide light beam to a narrow aperture.

Absorption enhancement can be achieved by utilizing the strong local field concentration of plasmons at the resonant wavelength of the nanostructures. Another important property of plasmons is their ability to generate energetic or "hot" electron–hole pairs by plasmon decay, which enabled the light detection with below bandgap photon illumination. The incident light can couple into surface plasmons by nanoantennas and the nonradiative decay of the plasmons results in hot electrons that can transfer across the Schottky barrier at the metal–semiconductor interface and can be detected as a photocurrent [16–18]. Thus, plasmonic effects hold great promise in enhancing the absorption, responsivity, and broadband spectrum in 2D material-based photodetectors.

8.2.3 Typical 2D-Based Infrared Photodetectors

8.2.3.1 Graphene Infrared Photodetectors

Graphene photodetectors can exhibit high bandwidth up to 500 GHz and ultrabroad spectrum because of the hot carrier-assisted electrical response and high carrier mobility. In 2010, T. Mueller and co-workers fabricated the metal–graphene–metal photodetectors where the multiple, interdigitated metal fingers are used, leading to the creation of a greatly enlarged, high E-field, light detection region [18]. As shown in Figure 8.3a, an asymmetric metallization scheme has been used to break the mirror symmetry of the built-in potential profile, allowing for the individual contributions to be summed to give the overall photocurrent. Figure 8.3b shows the gate-bias-dependent response with 1.55 μm light illumination, and the maximum photocurrent was observed at an optimized gate voltage of −15 V. The light power-dependent photocurrent is shown in Figure 8.3c, where a photocurrent of 15 mA is observed, corresponding to an external responsivity of 1.5 mA/W. Xia et al. also reported on ultrafast photocurrent response measurements in graphene-based FETs up to 40 GHz light intensity modulation frequencies, using a 1.55 μm excitation laser [19]. The ultrafast response in graphene-based photodetectors can be attributed to the dominated hot carrier transport [14]. D. Sun et al. used time-resolved scanning photocurrent microscopy to demonstrate that hot carriers, rather than phonons, dominate energy transport across a tunable graphene p–n junction excited by ultrafast laser pulses. The photocurrent response time varies from 1.5 ps at room temperature to 4 ps at 20 K, implying a fundamental bandwidth of ∼500 GHz [20].

In 2012, L. Vicarelli et al. fabricated the antenna-coupled graphene FETs (Figure 8.3d) and realized the terahertz detection. They demonstrated room temperature operation at 0.3 THz, showing that our devices can already be used in realistic settings, enabling large-area, fast imaging of macroscopic samples [21]. Although the fast speed and broad spectrum have been achieved, the reported responsivity was limited. Some approaches have then been proposed to enhance the responsivity. In 2013, X. Gan et al. demonstrated a waveguide-integrated graphene photodetector (as shown in Figure 8.3e) that simultaneously exhibits high responsivity, high speed, and broad spectral bandwidth. Figure 8.3f shows the light power-dependent responsivity, and the

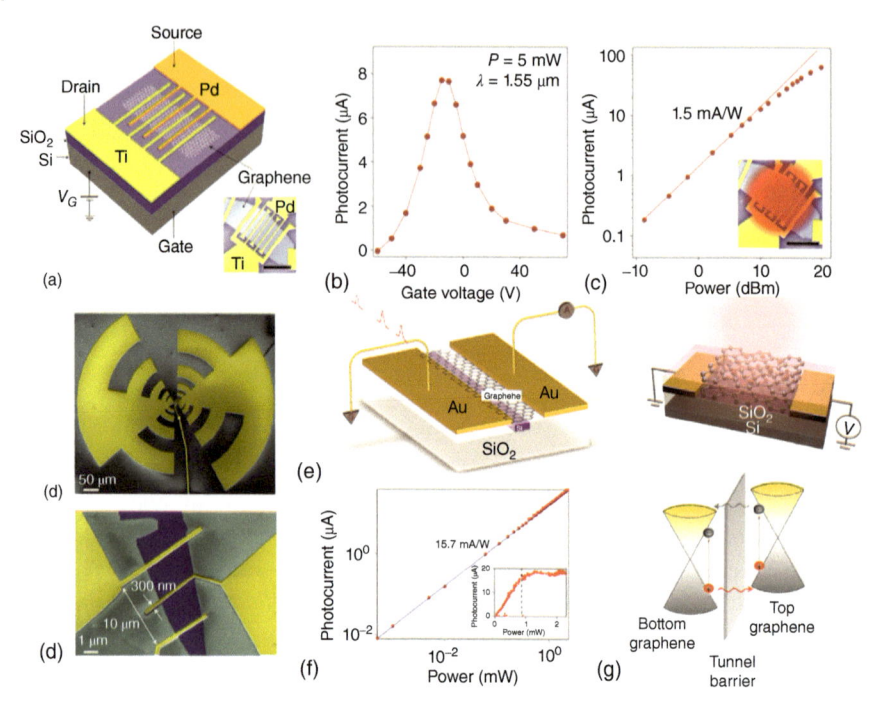

Figure 8.3 (a) Schematic diagram of graphene photodetectors with asymmetric interdigital electrodes. (b) Gate voltage-dependent photocurrent under 1.55 μm light illumination. (c) The photocurrent as a function of incident light power. Source: Reprinted with permission from Xia et al. [19]. Copyright 2010, Springer Nature. (d) Schematic diagram of the antenna-coupled graphene transistors. Source: Reprinted with permission from Gan et al. [20]. Copyright 2012, Springer Nature. (e) Schematic of the waveguide-integrated graphene photodetector. (f) Light power-dependent photocurrent, showing a R of 15.7 mA/W. Source: Reprinted with permission from Liu et al. [21]. Copyright 2013, Springer Nature. (g) Schematic of phototransistors composed of a pair of stacked graphene layers sandwiching a thin tunnel barrier (5 nm Ta_2O_5). Source: Reprinted with permission from Zhang et al. [22]. Copyright 2014, Springer Nature.

detector achieves a photoresponsivity of up to 15.7 mA/W [22]. C.-H. Liu and co-workers fabricated the phototransistors that comprise a pair of stacked graphene layers (top gate layer and bottom channel layer) sandwiching a thin tunnel barrier (5 nm Ta_2O_5) (Figure 8.3g). This device can exhibit the responsivity up to 1000 A/W in visible and 1–4 A/W from near-infrared (NIR) (1.3 μm) to mid-wave infra-red (MWIR) range (3.2 μm) with a bandwidth of 10–1000 Hz, rivaling state-of-the-art mid-infrared detectors without the need for cryogenic cooling [23]. Zhang et al. also reported high responsivity of 8.6 A/W compared to pure graphene phototransistors from the visible (532 nm) up to the mid-infrared (∼10 μm) in a single pure graphene photodetector, by introducing electron trapping centers and by creating a bandgap in graphene through band structure engineering [24].

8.3 2D Photodetectors with Sensitizers

8.3.1 Graphene-based Hybrids Detectors

In the pure 2D materials such as graphene photodetectors, the R is limited below 1 A/W because of the absence of gain. New device structures with photoactive sensitizers introducing a significant photogain for higher sensitivity have been proposed. Graphene showed the promising applications in broadband and ultrafast photodetectors; however, the absorption of the incident light is only 2.3% because of the monolayer thickness. On the other hand, the 2D TMDs possess strong optical absorption (1×10^7 m^{-1}) and sizable bandgap, which can complement the disadvantage of graphene. Roy et al. fabricated the graphene-on-MoS$_2$ binary heterostructures and demonstrated the highly sensitive photodetection [25]. Under light illumination, the MoS$_2$ layers can absorb the incident light and generate the electron–hole pairs, which can be separated at the interface, giving rise to the photogating effect. As a result, a very large photoconductive gain of $\sim 4 \times 10^{10}$ electrons per single photon and high responsivity of 5×10^8 A/W at room temperature can be achieved. The same device structures were also explored with CVD-grown graphene, and MoS$_2$ was also reported afterward [26]. In spite of the high gain and R in the TMD-sensitized graphene photodetectors, the response is very slow and the spectral coverage is limited in a visible range because of the large bandgap of TMDs.

To further improve the response speed and extend the spectral coverage, the colloidal quantum dots (CQDs) with a small bandgap such as PbS were firstly developed by G. Konstantatos et al. to sensitize graphene [27]. It was reported that CQDs have strong light absorption, broad absorption range from ultraviolet to short-wave infrared, and size-tunable bandgap through quantum confinement effect [28]. Figure 8.4a shows the concept of this type of detectors, consisting of graphene covered with a thin film of CQDs. Because of the high mobility of graphene and strong light absorption of CQDs, the hybrid devices exhibited high EQE of 25%, high responsivity up to 10^7 A/W, and broad coverage up to 1.6 µm [27]. Furthermore, the Konstantatos group developed the sensitizer layer from an electrically passive one to an active one, which has led to even better performance. This concept is shown in Figure 8.4b, in which the PbS layer is connected with a top electrode forming a Schottky photodiode [29]. As a result, the charge transfer efficiency from the sensitizer to the graphene was improved, leading to an EQE of 80% and the linear dynamic range increased to 100 dB. Recently, they further integrated the graphene-QD hybrid photodetectors with CMOS read-out circuits and demonstrated a broadband CMOS-based digital camera capable of capturing both visible and infrared light [30]. Figure 8.4c shows the schematic diagram of CMOS integration of graphene with 388×288 pixel image sensor read-out circuit. The specific D^* of 10^{12} Jones, fast speed of 0.1–1 ms, and the spectral range of 300–2000 nm were achieved. The successful integration of 120 000 pixel photodetectors in a single focal plane array has enabled high-resolution imaging showing the huge applications in night vision, food inspection, and medical imaging (Figure 8.4d).

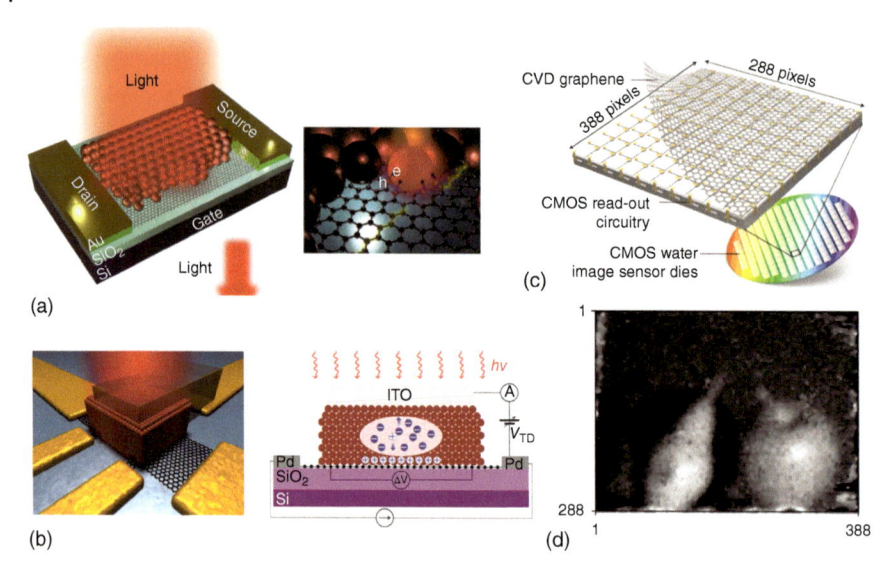

Figure 8.4 (a) Schematic of graphene CQD hybrid photodetectors, where graphene acts as the carrier transport channel and CQDs act as the strong light absorption layers. Source: Reprinted with permission from Konstantatos et al. [27]. Copyright 2014, Springer Nature. (b) Schematic of graphene–PbS hybrid photodetector integrated with a top transparent electrode. Source: Reprinted with permission from Nikitskiy et al. [29]. Copyright 2016, Springer Nature. (c) Computer-rendered impression of the CVD graphene transfer process on a single die containing an image sensor read-out circuit that consists of 388 × 288 pixels. Source: Reprinted with permission from Goossens et al. [30]. Copyright 2017, Springer Nature. (d) Near-infrared (NIR) and short-wave infrared (SWIR) light photograph of an apple and pear. Source: Reprinted with permission from Lee et al. [31]. Copyright 2015, American Chemical Society.

8.3.2 TMD-Based Hybrid Detectors

Instead of graphene, 2D TMDs were also chosen as the transport channel and sensitized with CQDs or other photoactive materials. The use of semiconducting 2D TMD channels can offer lower dark current, lower power consumption, and higher sensitivity through applying appropriate gate voltage. Photodetectors based on a diversity of 2D TMDs with monolayer or multilayer structures have been reported. By optimizing the device structures, the performances were much improved. For example, Lee et al. used a conducting NiO_x as gate electrodes directly on the MoS_2 channel without an insulator between them [31], this device exhibited high mobility of 500–1200 cm^2/V s, and the responsivity has reached as high as 5000 A/W with a fast response time of 2 ms. On the other hand, the spectral coverage was limited because of the sizable bandgap of 2D TMDs.

To further extend the spectrum and improve the performance, Kufer et al. reported the MoS_2–PbS hybrid photodetector and demonstrated a large gain of 10^6, a high responsivity of up to 10^6 A/W, and a D^* of 10^{12} Jones with extended spectral sensitivity up to 1.5 μm, determined by the absorption of the PbS CQDs [32]. Afterward, Huo et al. developed the devices using HgTe CQDs as sensitizers further extending the spectral coverage beyond 2 μm [33]. That is promising because HgSe and HgTe CQDs have been considered as low-cost route for mid-IR and far-IR detection because of their tunable bandgap throughout the full infrared spectrum with favorable optical properties. Interestingly, a thin TiO_2 buffer layer between MoS_2 and CQDs was implemented as a mediate interface (Figure 8.5a), which can preserve the gate modulation of current in MoS_2 allowing lower dark current and higher sensitivity. The D^* at a wavelength of 2 μm is close to 10^{12} Jones (Figure 8.5b), which was 2 orders of magnitude higher than prior reports from HgTe-based photodetectors as well as existing commercially available technologies based on extended InGaAs, InAs, or HgCdTe that also require thermoelectric cooling, demonstrating the great potential of hybrid 2D/QD detector technology in mid-IR applications with compelling sensitivity. Huo and Konstantatos also reviewed the recent progress and future prospects of 2D-based photodetectors [36].

Recently, Huo and Konstantatos developed a novel architecture composed of all 2D MoS_2 as shown in Figure 8.5c [34]. In this device, the out-of-plane MoS_2 homojunctions act as the sensitizing layer for the in-plane MoS_2 channel. $AuCl_3$ solution was used to p-type dope the top layers of MoS_2 while bottom layers remained n-type; thus, an out-of-plane p–n homojunction was formed at the interface that can serve as a sensitizing scheme and separate the photoexcited carriers efficiently. As a result, a high photogain of $>10^5$ and sensitivity up to 3.5×10^{14} Jones was achieved (Figure 8.5d). The same detector concept can be applied in other 2D semiconductors with much improved performance, particularly for those possessing low bandgap such as bP that can extend the spectral coverage of the 2D detectors into the mid-infrared range.

8.3.3 Plasmonic Sensitized Detectors

The strong local field concentration of plasmons at the resonant wavelength of the nanostructures can lead to the intense and resonant absorption enhancement. Based on this principle, the 2D photodetectors sensitized by metallic nanostructures have been demonstrated with improved photocurrent and extended spectral coverage. For example, Graphene nanodisks and Au plasmonic nanoantennas have also been reported to enhance the absorption efficiency in graphene from less than 3–30% [37, 38]. The plasmonic sensitizers can also generate the hot carriers enabling the photocurrent generation from photons with energy below the bandgap of the 2D channels. Hot-electron-induced photodetection has been reported in graphene [35] and MoS_2 [39]. A bilayer MoS_2 sensitized with a plasmonic antenna array also exhibited sub-bandgap photocurrent with a photogain of 10^5 and responsivity of 5.2 A/W at 1070 nm [39]. Similarly, the photocurrent in a graphene–antenna sandwich photodetectors (Figure 8.5e,f) was also enhanced by 800% because of the hot electron transfer and direct plasmon-enhanced excitation of intrinsic graphene electrons [35].

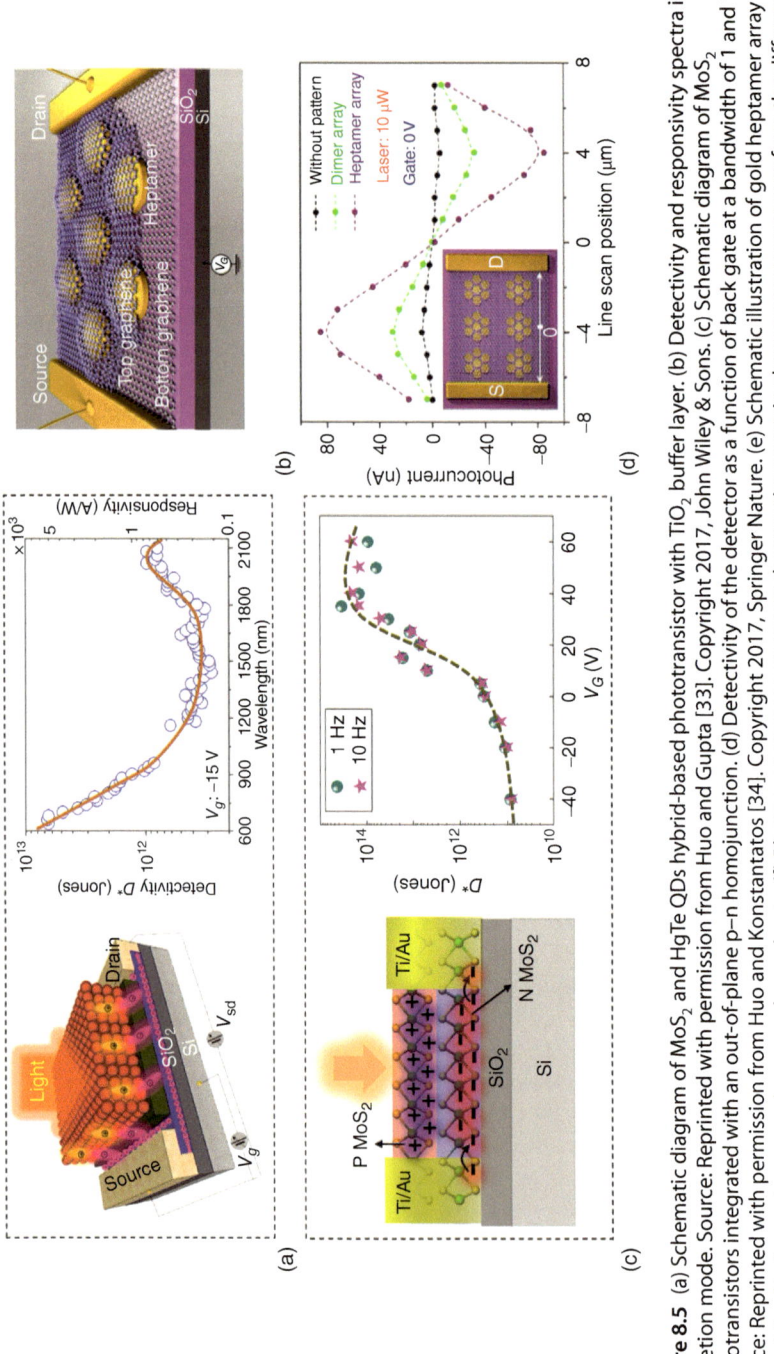

Figure 8.5 (a) Schematic diagram of MoS$_2$ and HgTe QDs hybrid-based phototransistor with TiO$_2$ buffer layer. (b) Detectivity and responsivity spectra in the depletion mode. Source: Reprinted with permission from Huo and Gupta [33]. Copyright 2017, John Wiley & Sons. (c) Schematic diagram of MoS$_2$ phototransistors integrated with an out-of-plane p–n homojunction. (d) Detectivity of the detector as a function of back gate at a bandwidth of 1 and 10 Hz. Source: Reprinted with permission from Huo and Konstantatos [34]. Copyright 2017, Springer Nature. (e) Schematic illustration of gold heptamer array sandwiched between two monolayer graphene sheets. (f) Photocurrent measurements show antisymmetric photocurrent responses from the different regions of the device corresponding to specific plasmonic antenna geometries, obtained along the line scan direction in inset. Source: Reprinted with permission from Fang et al. [35]. Copyright 2012, American Chemical Society.

8.4 New Infrared Photodetectors with Narrow Bandgap 2D Semiconductors

Recently, new monoatomic buckled crystals such as black phosphorene [4], silicene [40], germanene [41], etc., have been developed, which can offer a smaller bandgap of 0.2–2 eV, high mobility of 100–1000 cm^2/V s, and the possibility to serve as high-performance short- and mid-infrared photodetectors. In 2014, L. Li et al. fabricated the FETs based on few-layer black phosphorus crystals with thickness down to a few nanometers for the first time. The charge carrier mobility is found to be thickness-dependent, with the highest values up to ∼1000 cm^2/V s obtained for a thickness of ∼10 nm [4]. Afterward, Q. Guo et al. reported the bP mid-infrared detectors at 3.39 μm with high internal gain, resulting in an external responsivity of 82 A/W (Figure 8.6a) [42]. Noise measurements show that such bP photodetectors are capable of sensing mid-infrared light in the picowatt range, indicating a promising alternative material in mid-infrared wavelength range. M. Amani et al. developed the black phosphorus–arsenic alloys (b-PAs) for the infrared detection applications. The cutoff wavelength can be tuned from 3.9 to 4.6 μm and the detectivity can be on the order of 10^{10} Jones [46]. M. Long et al. also reported the black arsenic phosphorus-based long-wavelength infrared photodetectors, with room temperature operation up to 8.2 μm. They further fabricated the b-AsP/MoS$_2$ heterostructures to suppress the $1/f$ noise, and the specific detectivity is improved to be 4.9×10^9 Jones in

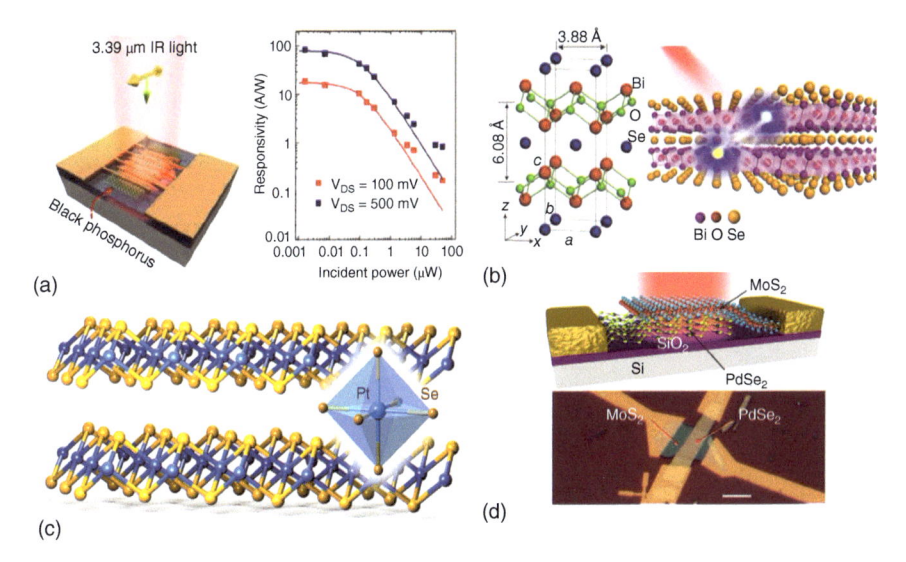

Figure 8.6 (a) Schematic of bP-based photodetectors. The right panel is the light power-dependent responsivity. Source: Reprinted with permission from Guo et al. [42]. Copyright 2016, American Chemical Society. (b) Atomic structures of 2D semiconducting Bi$_2$O$_2$Se. Source: Reprinted with permission from Yin et al. [43]. Copyright 2018, Springer Nature. (c) Atomic structure of PtSe$_2$. Source: Reprinted with permission from Yu et al. [44]. Copyright 2018, Springer Nature. (d) Schematic of MoS$_2$–PdSe$_2$ heterojunction-based devices. Source: Reprinted with permission from Long et al. [45]. Copyright 2019, American Chemical Society.

the 3–5 μm range [47]. Recently, H. Peng and co-workers developed the new type of 2D semiconducting Bi_2O_2Se that exhibits an ultrahigh Hall mobility value of >20 000 cm^2/V s at low temperatures. The top-gated FETs based on Bi_2O_2Se crystals down to the bilayer limit exhibit high Hall mobility values (up to 450 cm^2/V s), large current on/off ratios (>10^6), and near-ideal subthreshold swing values (~65 mV/dec) at room temperature [48]. They then demonstrated the infrared photodetector based on the 2D Bi_2O_2Se crystal, the high sensitivity of 65 AW^{-1} at 1200 nm, and ultrafast photoresponse of ~1 ps at room temperature was achieved (Figure 8.6b) [43]. From 2018, the noble metal dichalcogenides such as $PtSe_2$ and $PdSe_2$ were also demonstrated to perform excellent broadband mid-infrared detection. X. Yu et al. reported that bilayer $PtSe_2$ combined with defect modulation possesses strong light absorption in the mid-infrared region and realized a mid-infrared photoconductive detector operating in a broadband mid-infrared range, the atomic structures are shown in Figure 8.6c [44]. Afterward, M. Long et al. demonstrated the $PdSe_2$ phototransistors and their heterostructure (Figure 8.6d), which exhibit highly sensitive, air-stable, and operable long-wavelength infrared photodetections at room temperature. A high photoresponsivity of ~42.1 AW^{-1} (at 10.6 μm) was observed, which is an order of magnitude higher than the current record of platinum diselenide [45].

8.5 Future Outlook

In this section, we have summarized concisely the figure of merits, photodetection mechanism, and recent progress in the field of 2D material-based infrared photodetectors. The graphene-based photodetectors can enable the broadband detection from MWIR to submillimeter wavelength range with ultrafast speed because of the dominated hot carrier transport and high mobility. Another is a hybrid-based photodetector producing photogain mechanism by combining 2D materials with CQDs, perovskite, or plasmonic nanostructures. Benefiting from the huge gain, the hybrid devices exhibit extremely high sensitivity with video-imaging speed that is on a par with existing technologies. The recent emerging new 2D materials with narrow bandgap such as bP, Bi_2O_2Se, and $PdSe_2$ have also been the promising candidates in infrared applications. All the 2D materials possess the advantages of low cost, easy processing, flexibility, and integrations of silicon technologies, which enable the rapid development in next generation of infrared photodetectors.

8.5.1 Optoelectronic Memory of 2D Semiconductors

Nonvolatile memory devices play an important role in the modern communication and information industry [49]. However, so far, memory operation mainly utilizes the electrical or magnetic principle. Light, as a noninvasive stimulus [50], endows an intelligent strategy for logic, optoelectronic demodulator [51, 52] and imaging [53]. The optical manipulation on the charge memory [53–55] paves the way toward smart electronics [50, 54], artificial synapse, and neurons [56, 57]. Recently, 2D materials have shown interesting optoelectronic storage behavior [58]. Figure 8.7a shows the optoelectronic memory phenomenon in

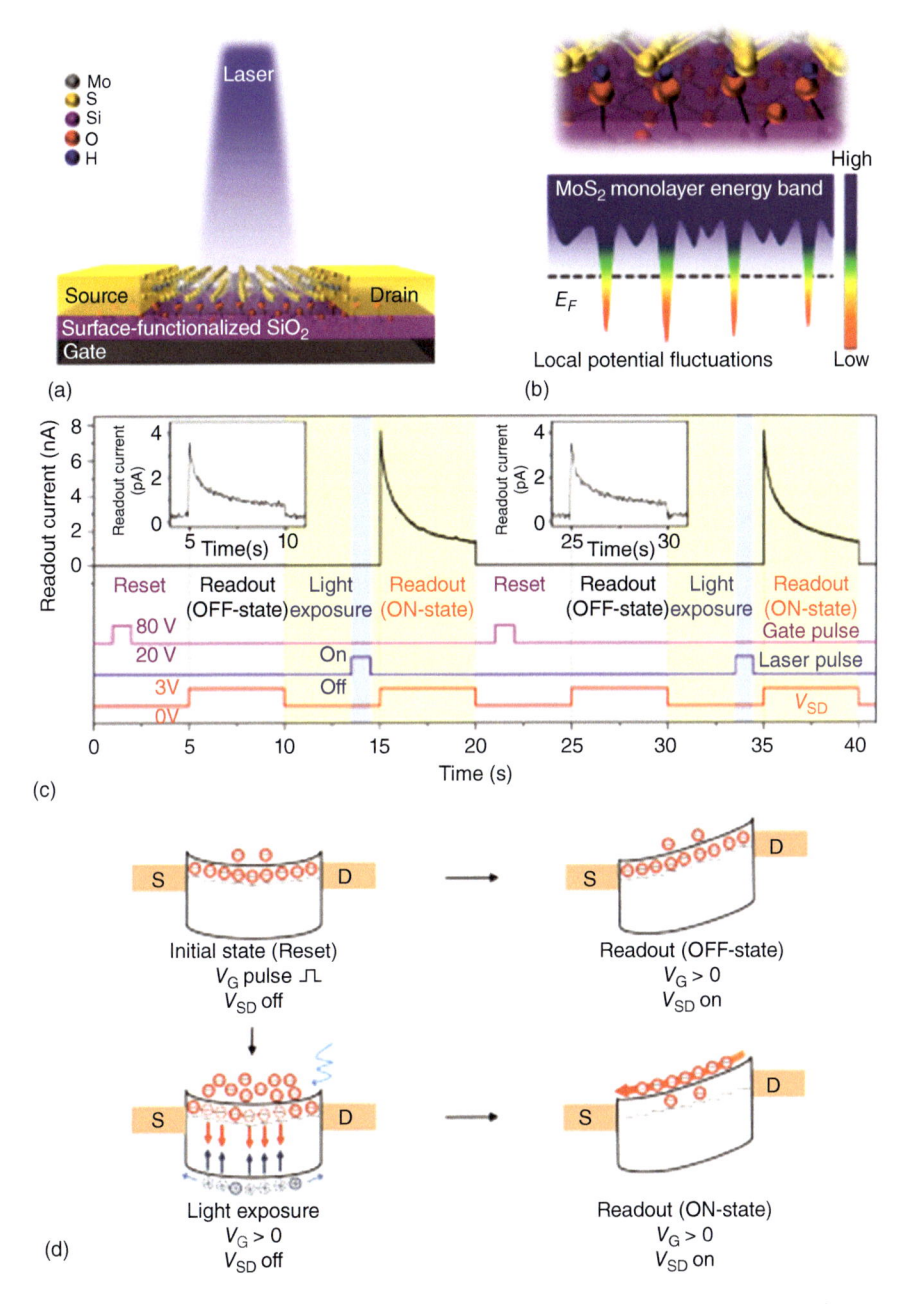

Figure 8.7 Optoelectronic memory devices based on monolayer MoS_2. (a) Schematic diagram of monolayer MoS_2 optoelectronic memory devices. (b) Energy band structure with local fluctuation potentials caused by functional silanol groups. (c) Readout current versus time. The gate pulses, laser pulses, and source–drain voltages are applied. (d) Schematic diagrams of the energy band and charge carrier at different work statuses. Source: Reprinted with permission from Lee et al. [58]. Copyright 2017, Springer Nature.

monolayer MoS$_2$. The oxygen plasma treatment creates the functional silanol groups (Si–OH) that modulate the energy band. The silanol group leads to the local potential fluctuations which trap the electrons (Figure 8.7b). The positive gate voltage pulse fills the trap states. The laser pulse produces the electron–hole pairs. The trapped electrons are recombined by holes. In addition, the photoexcited electrons increase the conductivity. The photoconductivity is persistent until the gate voltage pulses are applied as shown in Figure 8.7c and d.

Van der Waals heterostructure that cooperates the function of different 2D materials presents promising applications in optical memory. The graphene/MoS$_2$ van der Waals heterostructure has displayed the persistent photoconductivity as shown in Figure 8.8 [25]. The gate voltage is important to generate the potential well of holes in MoS$_2$. As shown in Figure 8.8d, as $V_g \ll V_T$, graphene is heavily hole-doped. The LED illumination (635 nm) generates the electron and hole pairs in MoS$_2$. The band bending forces the electrons into the graphene. The holes are then recombined, which results in the increases of resistance of transport channel. As we can see in Figure 8.8b, the photocurrent (I_p) decreases as the LED is on. With the switching off of LED, the I_p decreases. However, I_p did not recover its original states. In addition, I_p saturates at a smaller I_{off}, suggesting a long-term persistent photoconductivity. The I_p completely recovers its original state only by a gate voltage ($V_g > V_T$).

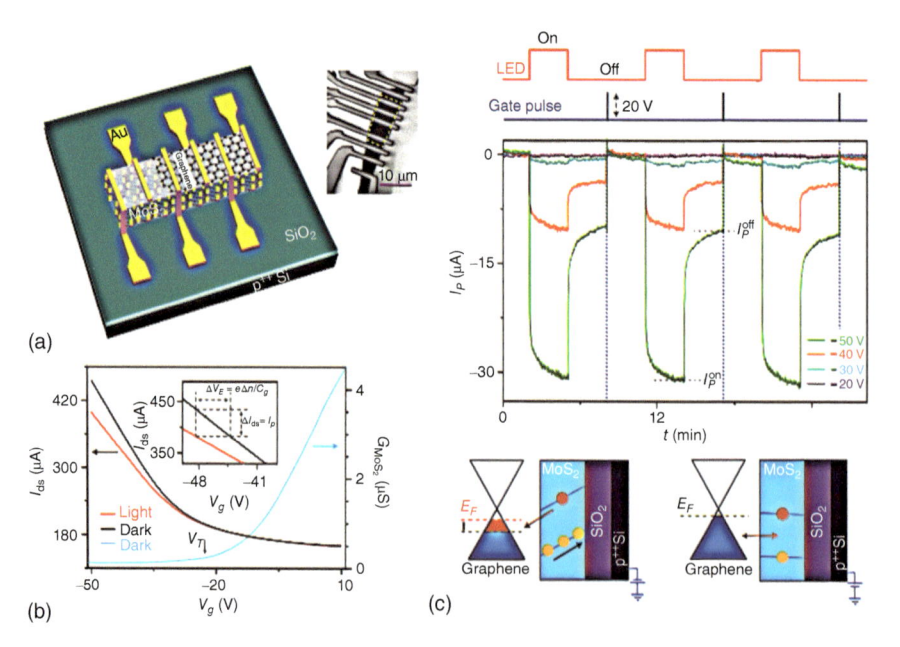

Figure 8.8 Persistent photoconductivity in graphene/MoS$_2$ heterostructure. (a) Schematic of the devices. (b) Transfer characteristic curves. (c) Photocurrent (I_p) as the function of time. LED pulses and gate voltage pulses are applied. The measurements are measured at different gate voltages (−20, −30, −40, and −50 V). (d) Energy band alignments showing the electron transferring and distribution. Source: Reprinted with permission from Roy et al. [25]. Copyright 2013, Springer Nature.

This gate voltage elevates the Fermi level, which suppresses the unidirectional transfer of electrons and equilibrates the charge distribution. However, M. D. Tran et al. reported a two-terminal multibit optical memory in the van der Waals heterostructure of MoS$_2$/h-BN/graphene [59]. As presented in Figure 8.9, graphene is utilized as the charge-trapping layer. The large negative source–drain voltage ($V_{sd} = -10\,\text{V}$) forces the electrons into the graphene

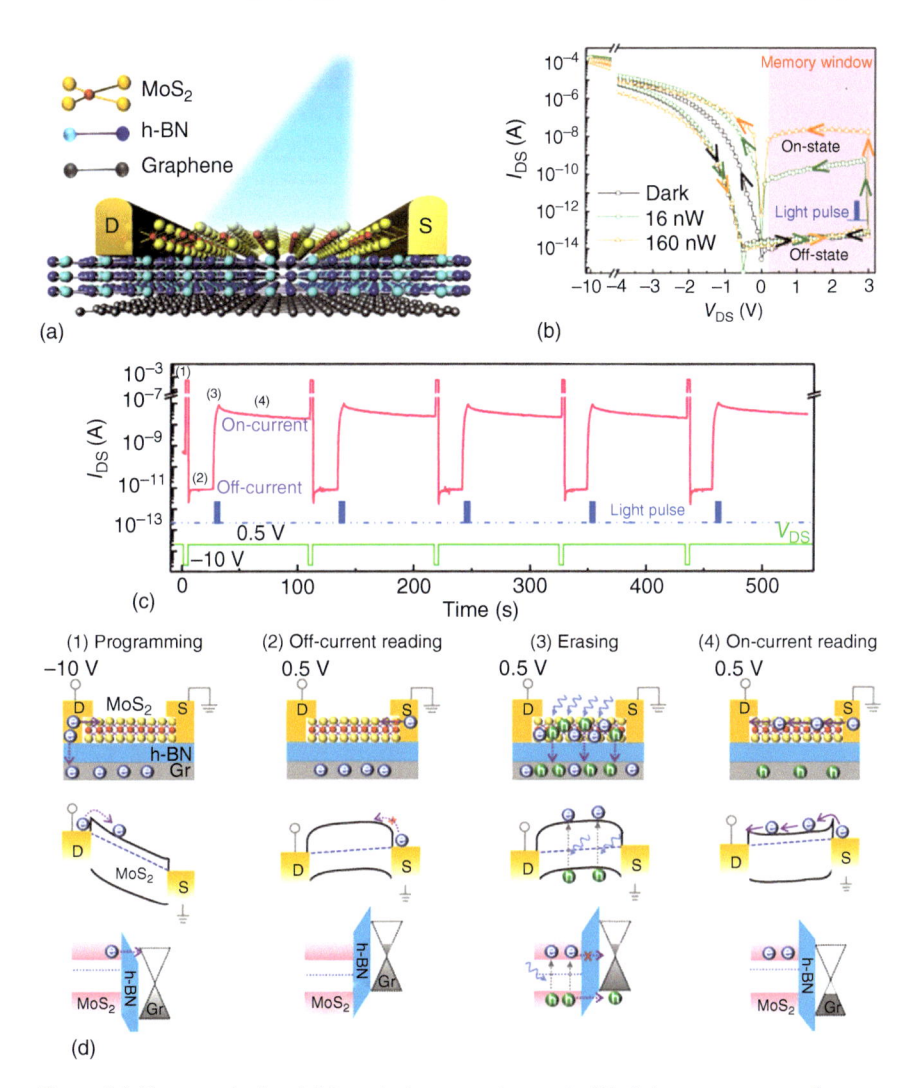

Figure 8.9 Two-terminal multibit optical memory in van der Waals heterostructure of MoS$_2$/h-BN/graphene. (a) Schematic of MoS$_2$/h-BN/graphene heterostructure. The MoS$_2$ is the semiconductive transport layer. The h-BN is the tunneling layer and blocking layer. The graphene is utilized as the charge-trapping layer. (b) I_{sd} versus V_{sd} curves that show the memory window. (c) Time-traced I_{sd} under application of light pulses and V_{sd} pulses. (d) Optoelectronic memory mechanism in MoS$_2$/h-BN/graphene heterostructure with device structure and energy band structure. Source: Reprinted with permission from Tran et al. [59]. Copyright 2019, John Wiley & Sons.

layer. A small $V_{sd} = 0.5$ V generates the small current (off-current). When the laser pulse applies, the photoexcited holes tunnel into a graphene layer and recombine with electrons. The photoexcited electrons in MoS$_2$ contribute to the on-current. The triangle barrier in the interface of MoS$_2$/h-BN protects the trapped electrons from recombination. The 18 memory states are obtained in this van der Waals heterostructure, which could store more than four-bit information.

The infrared spectra [60] are an important communication medium in night vision, military communication, object inspection, and medical diagnosis. Because of their low energy dissipation in the optical fiber, 850, 1310, and 1550 nm are typically utilized as the optical communication wavebands [61]. Therefore, the optoelectronic devices that would convert and store infrared information are significant for performing the optical data processing. Recently, Q. Wang et al. reported an infrared memory device using MoS$_2$/PbS van der Waals heterostructure [62]. As shown in Figure 8.10a, PbS nanoplates were grown on the surface of MoS$_2$ by CVD. PbS nanoplates are used as the infrared absorption layer, and MoS$_2$ is utilized as the charge transport layer. The special alignment of band structure leads to the localization of photoexcited holes as shown in Figure 8.10b. The localized holes serve as the positive voltage gate, which induces electrons in the MoS$_2$ transport layer. This is the physical

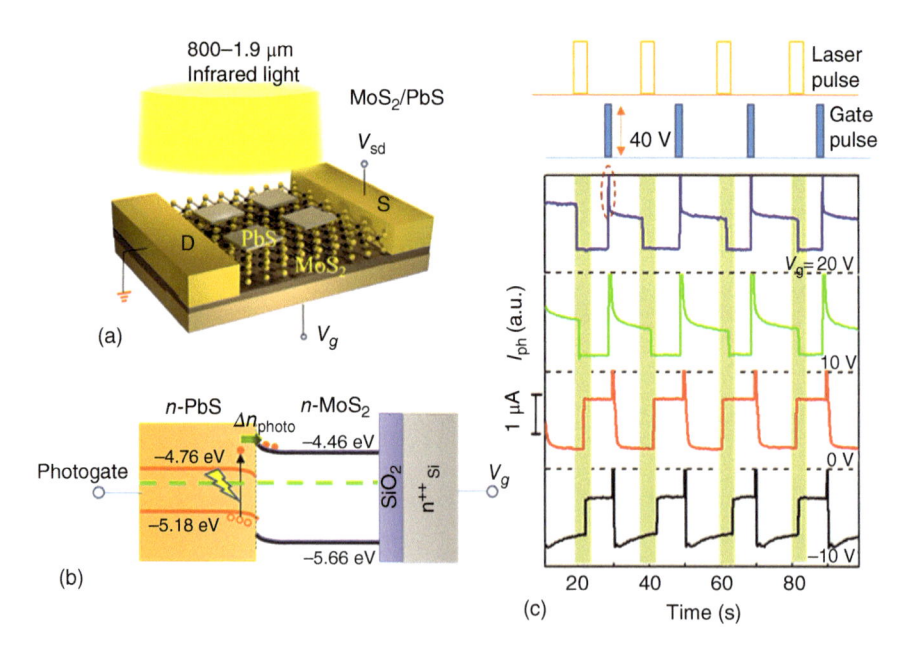

Figure 8.10 Nonvolatile infrared memory in PbS/MoS$_2$ heterostructure. (a) Device structure of PbS/MoS$_2$ heterostructure. (b) Energy band diagram showing the generation, transport, and recombination of photoexcited electron–hole pairs. (c) Experimental results of nonvolatile infrared memory with infrared pulse writing and gate voltage pulse erasing. The measurements were carried out at four background gate voltages (−10, 0, 10, and 20 V). Source: Reprinted with permission from Wang et al. [62]. Copyright 2018, American Association for the Advancement of Science.

origination of infrared memory. The trapped charges are removed by positive gate voltage pulses, which lift the Fermi level. The conduction band of MoS$_2$ further bends downward. The triangle potential barrier of PbS/MoS$_2$ becomes thinner. As a result, the electrons tunnel into the PbS conduction band and recombine with the photoexcited holes. As presented in Figure 8.10c, the positive V_g pulses (40 V) sharply increase the photocurrent because of the gate field effect. In addition, the persistent photocurrent goes back to its original state because of the recombination between tunneled electrons and localized holes.

The optical memory combines the function of light sensing and storage together, which shows great potential in the application of artificial visual system [63]. As compared with electrical methods, the optical memory benefits faster response and lower energy cost. Optical memory also allows the remote manipulation on the data coding and information processing. Further, the optical memory with multibit information storage may lead to brain-like computing. However, the optical memory has received far less investigation as compared with electric and magnetic routes. In the future, the room temperature, multistates and high-performance optical memory devices are strongly expected.

8.5.2 Solar Cells

The photovoltaic effect is an important physical phenomenon. The photovoltaic effect can induce the formation of spontaneous voltage in the devices. The application of photovoltaic effect is an important development direction of clean energy technology. The photovoltaic effect was first used to convert solar energy into electricity in 1954, based on a silicon p–n junction [64]. Currently, silicon-based devices are still the main choice for solar cells. The photovoltaic effect is based on the excitation of electrons and holes under illumination. Another important process is the separation of electron–hole pairs, thus effectively generating photovoltage. Because the bandgap of two-dimensional metal chalcogenides covers the spectrum from near-infrared to visible light, two-dimensional metal chalcogenides can effectively absorb light of all wavelengths [65]. In addition, the van der Waals junction without lattice mismatch is beneficial to reduce interface defects and accelerate electron–hole pair recombination. Therefore, two-dimensional metal chalcogenides provide a new material of choice for new ultralight and ultrathin solar cells.

The type II heterojunction has good band alignment and can separate the electron and hole excited by light. At present, researchers have studied the application of different type II heterojunctions with different two-dimensional metal chalcogenides in solar cells theoretically and experimentally [66–70]. For example, MoS$_2$/WS$_2$, MoSe$_2$/WSe$_2$, and ZrS$_3$/HfS$_3$ have been theoretically calculated [71, 72]. Among them, the ZrS$_3$/HfS$_3$ heterojunction has the highest energy conversion efficiency, which is due to the self-selected orbital splitting and exciton effect enhancement caused by symmetry destruction [73].

The light intensity and interlayer recombination will affect the short-circuit current and open-circuit voltage, and the light intensity can also adjust the number of generated carriers. The carrier concentration will affect the recombination in the two-dimensional heterogeneous junction. With the increase of light intensity, the carrier concentration increases, and the composite enhancement of the photoinduced carrier finally achieves the dynamic balance between the generated carrier and the recombined carrier. Wong et al. explored the company device structures based on the MoS_2/WSe_2 system to improve the photovoltaic quantum efficiency, and an EQE exceeding 50% was realized [74].

References

1 Schulman, D.S., Arnold, A.J., and Das, S. (2018). *Chem. Soc. Rev.* 47: 3037.
2 Yuan, T., Buchanan, D.A., Wei, C. et al. (1997). *Proc. IEEE* 85: 486.
3 Skotnicki, T., Hutchby, J.A., Tsu-Jae, K. et al. (2005). *IEEE Circuits Devices Mag.* 21: 16.
4 Li, L., Yu, Y., Ye, G.J. et al. (2014). *Nat. Nanotechnol.* 9: 372.
5 Radisavljevic, B., Radenovic, A., Brivio, J. et al. (2011). *Nat. Nanotechnol.* 6: 147.
6 Ovchinnikov, D., Allain, A., Huang, Y.-S. et al. (2014). *ACS Nano* 8: 8174.
7 Sucharitakul, S., Goble, N.J., Kumar, U.R. et al. (2015). *Nano Lett.* 15: 3815.
8 Bandurin, D.A., Tyurnina, A.V., Yu, G.L. et al. (2017). *Nat. Nanotechnol.* 12: 223.
9 Mak, K.F., Lee, C., Hone, J. et al. (2010). *Phys. Rev. Lett.* 105: 136805.
10 Wen, W., Zhu, Y., Liu, X. et al. (2017). *Small* 13: 1603788.
11 Kaasbjerg, K., Thygesen, K.S., and Jacobsen, K.W. (2012). *Phys. Rev. B* 85: 115317.
12 Huo, N., Yang, Y., Wu, Y.-N. et al. (2018). *Nanoscale* 10: 15071.
13 Koppens, F.H.L., Mueller, T., Avouris, P. et al. (2014). *Nat. Nanotechnol.* 9: 780.
14 Gabor, N.M., Song, J.C.W., Ma, Q. et al. (2011). *Science* 334: 648.
15 Wu, J., Schmidt, H., Amara, K.K. et al. (2014). *Nano Lett.* 14: 2730.
16 Mubeen, S., Hernandez-Sosa, G., Moses, D. et al. (2011). *Nano Lett.* 11: 5548.
17 Knight, M.W., Sobhani, H., Nordlander, P., and Halas, N.J. (2011). *Science* 332: 702.
18 Mueller, T., Xia, F., and Avouris, P. (2010). *Nat. Photonics* 4: 297.
19 Xia, F., Mueller, T., Lin, Y.-M. et al. (2009). *Nat. Nanotechnol.* 4: 839.
20 Sun, D., Aivazian, G., Jones, A.M. et al. (2012). *Nat. Nanotechnol.* 7: 114.
21 Vicarelli, L., Vitiello, M.S., Coquillat, D. et al. (2012). *Nat. Mater.* 11: 865.
22 Gan, X., Shiue, R.-J., Gao, Y. et al. (2013). *Nat. Photonics* 7: 883.
23 Liu, C.-H., Chang, Y.-C., Norris, T.B., and Zhong, Z. (2014). *Nat. Nanotechnol.* 9: 273.
24 Zhang, Y., Liu, T., Meng, B. et al. (2013). *Nat. Commun.* 4: 1811.
25 Roy, K., Padmanabhan, M., Goswami, S. et al. (2013). *Nat. Nanotechnol.* 8: 826.
26 Zhang, W., Chuu, C.-P., Huang, J.-K. et al. (2014). *Sci. Rep UK.* 4: 3826.
27 Konstantatos, G., Badioli, M., Gaudreau, L. et al. (2012). *Nat. Nanotechnol.* 7: 363.

28 Konstantatos, G., Howard, I., Fischer, A. et al. (2006). *Nature* 442: 180.
29 Nikitskiy, I., Goossens, S., Kufer, D. et al. (2016). *Nat. Commun.* 7: 11954.
30 Goossens, S., Navickaite, G., Monasterio, C. et al. (2017). *Nat. Photonics* 11: 366.
31 Lee, H.S., Baik, S.S., Lee, K. et al. (2015). *ACS Nano* 9: 8312.
32 Kufer, D., Nikitskiy, I., Lasanta, T. et al. (2015). *Adv. Mater.* 27: 176.
33 Huo, N., Gupta, S., and Konstantatos, G. (2017). *Adv. Mater.* 29: 1606576.
34 Huo, N. and Konstantatos, G. (2017). *Nat. Commun.* 8: 572.
35 Fang, Z.Y., Liu, Z., Wang, Y.M. et al. (2012). *Nano Lett.* 12: 3808.
36 Huo, N. and Konstantatos, G. (2018). *Adv. Mater.* 30: 1801164.
37 Naik, G.V., Shalaev, V.M., and Boltasseva, A. (2013). *Adv. Mater.* 25: 3264.
38 Echtermeyer, T.J., Britnell, L., Jasnos, P.K. et al. (2011). *Nat. Commun.* 2: 458.
39 Wang, W., Klots, A., Prasai, D. et al. (2015). *Nano Lett.* 15: 7440.
40 Tao, L., Cinquanta, E., Chiappe, D. et al. (2015). *Nat. Nanotechnol.* 10: 227.
41 Ni, Z., Minamitani, E., Ando, Y., and Watanabe, S. (2017). *Phys. Rev. B* 96: 075427.
42 Guo, Q., Pospischil, A., Bhuiyan, M. et al. (2016). *Nano Lett.* 16: 4648.
43 Yin, J., Tan, Z., Hong, H. et al. (2018). *Nat. Commun.* 9: 3311.
44 Yu, X., Yu, P., Wu, D. et al. (2018). *Nat. Commun.* 9: 1545.
45 Long, M., Wang, Y., Wang, P. et al. (2019). *ACS Nano* 13: 2511.
46 Amani, M., Regan, E., Bullock, J. et al. (2017). *ACS Nano* 11: 11724.
47 Long, M., Gao, A., Wang, P. et al. (2017). *Sci. Adv.* 3: e1700589.
48 Wu, J., Yuan, H., Meng, M. et al. (2017). *Nat. Nanotechnol.* 12: 530.
49 Wong, H.S.P. and Salahuddin, S. (2015). *Nat. Nanotechnol.* 10: 191.
50 Gorostiza, P. and Isacoff, E.Y. (2008). *Science* 322: 395.
51 Tan, H., Liu, G., Yang, H. et al. (2017). *ACS Nano* 11: 11298.
52 Tan, H., Liu, G., Zhu, X. et al. (2015). *Adv. Mater.* 27: 2797.
53 Lei, S., Wen, F., Li, B. et al. (2014). *Nano Lett.* 15: 259.
54 Lv, Z., Wang, Y., Chen, Z. et al. (2018). *Adv. Sci.* 5: 1800714.
55 Liu, F., Zhu, C., You, L. et al. (2016). *Adv. Mater.* 28: 7768.
56 Maier, P., Hartmann, F., Emmerling, M. et al. (2016). *Phys. Rev. Appl.* 5: 054011.
57 Lee, M., Lee, W., Choi, S. et al. (2017). *Adv. Mater.* 29: 1700951.
58 Lee, J., Pak, S., Lee, Y.-W. et al. (2017). *Nat. Commun.* 8: 14734.
59 Tran, M.D., Kim, H., Kim, J.S. et al. (2019). *Adv. Mater.* 31: 1807075.
60 Rogalski, A. (2003). *Prog. Quantum Electron.* 27: 59.
61 Dutton, H.J. (1998). *Understanding Optical Communications*. Upper Saddle River, NJ: Prentice Hall PTR.
62 Wang, Q., Wen, Y., Cai, K. et al. (2018). *Sci. Adv.* 4: eaap7916.
63 Chen, S., Lou, Z., Chen, D., and Shen, G. (2018). *Adv. Mater.* 30: 1705400.
64 Chapin, D.M., Fuller, C.S., and Pearson, G.L. (1954). *J. Appl. Phys.* 25: 676.
65 Jariwala, D., Sangwan, V.K., Lauhon, L.J. et al. (2014). *ACS Nano* 8: 1102.
66 Tsai, M.-L., Li, M.-Y., Retamal, J.R.D. et al. (2017). *Adv. Mater.* 29: 1701168.
67 Furchi, M.M., Pospischil, A., Libisch, F. et al. (2014). *Nano Lett.* 14: 4785.
68 Pezeshki, A., Shokouh, S.H.H., Nazari, T. et al. (2016). *Adv. Mater.* 28: 3216.

69 Lee, C.-H., Lee, G.-H., van der Zande, A.M. et al. (2014). *Nat. Nanotechnol.* 9: 676.
70 Furchi, M.M., Höller, F., Dobusch, L. et al. (2018). *npj 2D Mater. Appl.* 2: 3.
71 Kang, J., Tongay, S., Zhou, J. et al. (2013). *Appl. Phys. Lett.* 102: 012111.
72 Zhao, Q., Guo, Y., Zhou, Y. et al. (2018). *Nanoscale* 10: 3547.
73 Britnell, L., Ribeiro, R.M., Eckmann, A. et al. (2013). *Science* 340: 1311.
74 Wong, J., Jariwala, D., Tagliabue, G. et al. (2017). *ACS Nano* 11: 7230.

9

Perspective and Outlook

In the past 15 years, the research about 2D materials has made great progress. Because of the ultrathin thickness of 2D materials, almost all atoms are exposed to the surface, so they are very sensitive to the external control, and can produce many unusual physical properties that bulk materials do not have. The basic processes of growth and evolution of 2D semiconductor materials require a deeper understanding to promote their applications in electronics, information, energy, environment, and biology.

In the controllable preparation of 2D materials, researchers have done a lot of excellent work; improved chemical vapor deposition (CVD), physical vapor deposition (PVD), metal organic chemical vapor deposition (MOCVD), and other methods have been used to grow the large-scale films. Although some promising results have been obtained, it is far from industrial application. To date, there has been a lack of effective methods for preparing large-area (wafer-scale) and high-quality samples (uniform thickness and low density of defects). 2D materials with wafer scale can be applied in flexible wearable devices, integrated circuits, optical components, etc. The preparation of wafer-scale and high-crystal quality 2D materials is one of the important challenges currently faced, and it is an urgent problem to be solved.

In 2D single materials, studies have found that there are thousands of 2D materials in nature. According to research studies, there are fewer 2D semiconductors with similar properties to commercial silicon. Very recently, 2D black phosphorus, black arsenic, and InSe exhibit high carrier mobility, which are expected to be applied to the future high-speed devices; however, they have serious environmental degradation problems, which directly limit their application in high-speed optoelectronic devices. Therefore, it is necessary to continuously explore the new 2D materials to provide support for the future applications of low-power, high-performance devices. Until now, unit and binary 2D materials have been extensively synthesized and investigated, and ternary and multivariate materials need further study.

In 2D doped materials, controllable synthesis of doping concentration and impurity atoms is a key point. The electrical and optoelectronic properties of 2D doped materials have been extensively investigated, and magnetism, catalysis, and thermoelectricity should be brilliant in 2D doped materials, but few studies have been performed.

Various 2D materials can form heterostructures because the lattice mismatch of 2D materials can form Moiré superlattices with novel physical properties, such as quantum Hall effect. Relevant insulation and superconductivity have been reported in twisted bilayer graphene Moiré superlattices. In addition, Moiré superlattices have been predicted to be the main excited states, such as Moiré exciton bands. The Moiré exciton band can be adjusted by the twist angle, temperature, and gate voltage of the heterostructure, which provides a new idea for exploring and controlling the excited states of matter. On the other hand, 2D heterostructures have many advantages, such as nonsuspended heterostructure interface, suitable band-order type, charge separation of ultrafast response, ultralong exciton lifetime, and high photoelectric conversion efficiency separation, which make them have great potential applications in the field of optical collection in the future, which need further exploration.

In theory, modeling plays an increasingly important role in the research of two-dimensional semiconductor materials. The specific performance is in two aspects: (i) to understand the experimental results in a more systematic and basic way, in order to verify the results of mutual discovery by theory and experiment and (ii) to discover new properties of materials and design new materials. At present, based on theoretical results, researchers have discovered many of the characteristics of 2D semiconductor materials and have been applied to practical work. In addition, scholars have found that the electronic properties of 2D semiconductor materials have driven the breadth of their applications. Different from traditional materials, the existing methods of size control, strain engineering, and power plant adjustment can adjust the physical properties of 2D materials. In addition, based on these three methods to obtain important information, it will play an important role in promoting the research and development of new nanomaterials and optoelectronic devices in the future.

Index

a

all-2D MoS$_2$ 153
alloying   99–121
α- and β-arsenene   67
α-As$_x$Sb$_y$ 103
α-graphyne   57
angle resolved photoemission spectroscopy (ARPES)   94
anisotropic phonon softening   41
anisotropic transport properties of 2D group-VA semiconductors   67–69
antenna-coupled graphene field-effect transistors   149
antibonding p$_x$–p$_y$ π* bands   14
antibonding p$_z$ π* bands   14
antiferromagnetic SnS$_2$ 121
antimonene   67, 123
armchair α-graphyne nanoribbons   37, 38
armchair graphene nanoribbon (AGNR)   35, 36
armchair ribbon (a-PNR)   40
armchair-shaped MX$_2$ nanoribbons   39
arsenene   67, 68, 123
As$_{Se}$–MoSe$_2$ 74
asymmetric ZGNRs   55, 56
atomic force microscopy (AFM)   91, 132
Au film assisted exfoliation method   81
Au plasmonic nanoantennas   153

b

Bader charge analysis   14
band-to-band tunneling (BTBT)   126
bandwidth   147, 149, 150
b-AsP/MoS$_2$ heterostructures   155
β-antimonene   67
β-As$_x$Sb$_y$ 103
β-bismuthene   68
bias-dependent transmission coefficient   55
biaxial strain   41, 43
bilayer graphene   24, 26, 31–33, 166
bilayer MoS$_2$ 153
Bi$_2$S$_3$ 84
bismuthene   67, 68
Bi$_2$Te$_3$/Sb$_2$Te$_3$ heterostructures   135
black arsenic   93, 155, 165
black phosphorus   39, 51, 83, 84, 123, 126, 143, 144, 155, 165
black phosphorus-arsenic alloys (b-PAs)   155
black phosphorus (Bp)-based FETs   143
bolometric effect   147–148
Boltzmann constant   67
Boltzmann transport equation   64
boron nitride (BN)   4, 37, 123, 144
Bose–Einstein statistics   62
bp based photodetectors   155
bp mid-infrared detectors   155
Bp/MoS$_2$ heterojunction diode   126
BP-on-WSe$_2$ 126
BP-on-WSe$_2$ photodetectors   128
Brillouin zone center   41
Brillouin zone folding   28
broadband and ultrafast photodetectors   151
Br$_{Se}$-MoSe$_2$ 74

Two-Dimensional Semiconductors: Synthesis, Physical Properties and Applications,
First Edition. Jingbo Li, Zhongming Wei, and Jun Kang.
© 2020 Wiley-VCH Verlag GmbH & Co. KGaA. Published 2020 by Wiley-VCH Verlag GmbH & Co. KGaA.

bulk 2D crystal 81
bulk GaS 49

c

carbon–carbon bonds 14
carrier–phonon scattering rate 64
carrier Schottky barriers 69–70, 72
carrier type modulation 103–104
$CdS_xSe_{(1-x)}$ alloys 102, 113
chalcogenide atoms 63
charge-carrier mobility 143, 155
charge transport properties
    anisotropic transport properties of 2D group-VA semiconductors 67–69
    phonon scattering mechanism of transition-metal dichalcogenides 63–67
    phonon scattering mechanisms of graphene 61–63
chemical methods 131–137
chemical vapor deposition (CVD) 79, 89, 101, 110, 113–114, 129, 145, 165
chemical vapor transport (CVT) 101, 110–111
colloidal quantum dots (CQDs) 151
$Co_{0.16}Mo_{0.84}S_2$ alloy 114
conduction band maximum (CBM) 14, 16
conduction band minimum (CBM) 35, 64
conduction band offset (CBO) 17
contacts between 2D semiconductors and metal electrodes
    carrier Schottky barriers 69–70
    partial Fermi level pinning and tunability 70–72
    role of defects in enhanced Fermi level pinning 72–75
controllable two-dimensional semiconductor materials 99
Coulomb interaction 10, 48
Coulomb potential 10
Coulomb scattering 145
Curie temperature 46, 129

d

decay time 146
density functional perturbation theory (DFPT) 62, 64
density functional theory (DFT) 9–10, 47, 62, 64, 102
deposition temperature 86, 111, 112
Desai's Au assisted exfoliation process 82
detectivity 147, 155
diagonal ribbon [d-PNR] 40
dielectric constant 48–50, 145
Dirac cone shifting 41
Dirac-fermion-like behavior 28
direct-indirect bandgap transition 43
doping 4, 41, 56, 99–121
d-orbitals 71
double resonance Raman scattering 41
dry-transfer method 130
dynamically screened Coulomb interaction 10

e

edge functionalization 40
edge morphology 40
electric field modulation 35, 48–51
electrocatalysts 96
electrochemistry 96
electron-and-hole radiation recombination efficiency 126
electron energy loss spectroscopy (EELS) 115
electron flux 146
electron-hole asymmetry 31, 32
electronic device 61, 126
electronic quasiparticle excitation 10
electronic structure of 2D semiconductors
    graphyne family members 11–14
    nitrogenated holey graphene 14–15
    transition metal dichalcogenides 15–19
electron injection efficiency 69
electron-phonon coupling strength 61
electron–phonon interaction matrix elements 62

electron–phonon mediated superconductivity 67
electron–phonon scattering 61
energy dispersive X-ray spectroscopy (EDS/EDX) 95
Everhart–Thornley detector 91
external coulomb scattering 145
external quantum efficiency (EQE) 129, 146
external strains 41, 46

f

Fe-doped SnS_2 single layer 4, 121
Fermi–Dirac distribution 55
Fermi level pinning (FLP) 43, 70–75, 109
Fermis golden rule 62, 64
ferromagnetic (FM) symmetric ZGNRs 61
field-effect transistors (FETs) 1, 70, 109, 123, 126, 129, 143–145
fixed lattice 105
folded spectrum method (FSM) 23
free energy change 133
free-standing graphene 41
funneling effect 47

g

$GaSe_{1-x}Te_x$ alloy 102
gate length 145
Geim, Ander 1
generalized gradient approximation (GGA) 10
germanene 123, 155
GeS_xSe_{1-x} alloys 103
giant Stark effect (GSE) 50
G/NHG heterostructures 27, 28, 31, 32
graphdiyne 42
graphdiyne (graphyne-2) 11
graphdiyne nanoribbons 37
grapheme 83, 123, 138
graphene 1
  Dirac band structure 1
  in-plane dielectric constant 48
  mobility of electrons 2
  out-of-plane dielectric constant 48
  phonon scattering mechanism 61
  room temperature quantum Hall effect 2–3
  single-molecule detection 2
  size control 35–40
  thermal conductivity 2
graphene-antenna sandwich photodetectors 153
graphene based hybrids detectors 151–152
graphene/BN heterostructures 27
graphene films 1–2
graphene/h-BN/TMDC/h-BN/graphene heterojunction 129
graphene/h-BN vertical heterojunction 126
graphene infrared photodetectors 149–150
graphene–MoS_2 137
graphene-MoS_2 van der Waals heterostructure 158
graphene nanodisks 153
graphene nanoribbons (GNRs) 35, 36, 48, 55–57
graphene/nitrogenated-holey-graphene (G/NHG) heterostructure 26
  band structure 28
  ordered stacking *versus* moiré pattern 26–30
  renormalized Fermi velocity 31–33
graphene-on-MoS_2 binary heterostructures 151
graphene-QD hybrid photodetectors 151
graphyne 37
graphyne-3 42
graphyne-4 42
graphyne family members 11–14
graphyne nanoribbon 57–59
Green's function 10, 11, 59
GW approximation 10

h

Hartree–Fock exchange 10
hBN-MoS_2 137, 138
h-BN substrates 135, 136
hexagonal boron nitride 6, 123, 144
$HfS_{2(1-x)}Se_{2x}$ 102, 113

HgTe CQDs 153
highest occupied molecular orbital (HOMO) 37
highly crystalline MoO$_3$ nanoribbons 88
high-mobility 4-inch wafer-scale films 89
high-performance field-effect transistors 123
high performance short-and mid-infrared photo-detectors 155
high-resolution transmission electron microscopy (HRTEM) 80, 84, 92–94
H lattice constants 105
hole induced potentials 33
homogeneous tensile strain 42
hot-electron-induced photodetection 153
2H-phase WSe$_{2(1-x)}$Te$_{2x}$ single-layer alloys 119
Huang's modified exfoliation process 81

i
ideal Mutated-junction 125
indirect-to-direct bandgap transition 38, 43, 46
infrared photodetectors
  bandwidth 147
  decay time 146
  detectivity 147
  electron flux 146
  external quantum efficiency 146
  graphene photodetectors 149
  internal quantum efficiency 146
  with narrow bandgap 2D semiconductors 155–156
  noise current 147
  noise equivalent power 147
  photodetection mechanism 147
  photo-gain 146
  photon flux 146
  responsivity 146
  rise time 146
  signal to noise ratio 147
  time constant 146
infrared spectrum 145, 153, 160
InSe 6, 83, 85, 144, 165
interface phonon scattering 145
interlayer binding energy 27, 28
interlayer coupling 1, 20, 22, 24, 30, 46, 69, 138, 139
inter-layer quantum coupling effect 125
internal defect 145
internal quantum efficiency (IQE) 146
isolated metal electrodes 71
isolated single-layer MoSe$_2$ 71

j
Janus monolayers 123

l
Landauer–Buttiker formula 59
large-scale and uniform thickness 2D semiconductors 85
large-scale MoS$_2$ thin layers 87
lateral heterostructures 125, 135
lateral WX$_2$/MoX$_2$ heterostructures 137
layered black phosphorus 39
linear GSE coefficient 50
linear scaling density functional theory (DFT) method 20
linear scaling three-dimensional fragment (LS3DF) method 10, 23
liquid method 129–131
liquid-phase exfoliation 79, 81–85
Li$_x$MoS$_2$ compound 83
local density approximation (LDA) 10, 102
longitudinal acoustic (LA) mode 63, 64
low-dimensional TMDs nanostructures 39
lowest unoccupied molecular orbital (LUMO) 37

m
magnetic force microscopy (MFM) 47
magneto-resistance (MR) 61

Matthiessen rule 66
maximum free energy change 133
mechanical exfoliation 1, 21, 39, 79–81, 85, 87, 102, 110, 111
mechanically exfoliated 2D material 35
metal-graphene-metal photodetectors 149
metal/MoSe$_2$ systems 72
metal-organic chemical vapour deposition technique (MOCVD) 89
metal Pd-semiconductor WSe$_2$ channel 108
metal-semiconductor contact 69, 109
metal-semiconductor phase transition phenomena 104
metastable/non-diffusion phase diagram 105
micro-Raman spectra 94
mini Dirac cones 19
Mn doping 107
Moiré exciton band 166
Moiré excitons 19
Moiré-patterned G/NHG heterostructures 31
Moiré superlattices 166
molybdenum trioxide (MoO$_3$) 88, 133
monoatomic buckled crystals 155
monolayer MoS$_2$ 42, 51, 80, 83, 89, 91, 94, 104, 135, 144, 145, 157
MoS$_2$ 4
MoSe$_2$ 80
MoSe$_{2-2x}$S$_{2x}$ 101
MoS$_2$/MoSe$_2$ heterostructures 19
MoS$_2$/MoSe$_2$/MoTe$_2$ 2D compounds 100
MoS$_{2(1-x)}$Se$_{2x}$ 4, 111, 113, 114
MoS$_2$ shape transformation phenomenon 91
MoS$_2$/WS$_2$ heterojunction arrays 126
MoS$_{2x}$Te$_{2(1-x)}$ 113
MoTe$_2$ 80
MoX$_2$-WX$_2$ lateral heterostructures 17
MX$_2$ 123
    monolayers 16, 17
MXenes 123, 124

n

narrow gap physical vapor deposition (NGPVD) method 112
NbSe$_2$/W$_x$Nb$_{1-x}$Se$_2$/WSe$_2$ heterojunction 109
NbSe$_2$/W$_x$Nb$_{1-x}$Se$_2$/WSe$_2$ van der Waals (M-vdW) junction 108
n-channel devices 143
near-infrared multi-band absorption 126
nitrides 123, 124
nitrogenated holey graphene (NHG) 11, 14–15
noise current 147
noise equivalent power (NEP) 147
non-equilibrium Green's function (NEGF) formalism 11, 59
nonvolatile memory devices 156
Novoselov, Konstantin 1–3
nucleation free energy change 133

o

one-body Green's function 10
one-dimensional carbon nanotubes 1
1D graphyne 37
one-dimensional nanoribbons 35
optical absorption and photoluminescence 94
optical memory 158, 159, 161
optoelectronic memory of 2D semiconductors 156–161
out-of-plane acoustic modes 64
out-of-plane A$_g$-like peak 47
out-of-plane MoS$_2$ homojunctions 153
out-of-plane phonons 62
oxides 91, 111, 123, 144

p

palsmonic sensitized detectors 153–154
PbS nanoplates 160
PDMS viscoelastic stamps 130
PdSe$_2$ phototransistors 156
perfect-MoSe$_2$ 74
perfect-MoSe$_2$/Au contact 75
perfect-MoSe$_2$/metal systems 75

perovskite/plasmonic nanostructures 156
phase change 104–107
phase change memory (PCM) 104
phonon occupation number 62
phonon scattering mechanisms 145
　of graphene 61–63
　of transition-metal dichalcogenides 63–67
photodetection mechanism
　bolometric effect 147–148
　photogating effect 148
　photo-thermoelectric effect 147
　photovoltaic effect 148
　plasmonic effect 148–149
photodetectors 123, 126–129, 145–156
photodiodes 146, 148, 151
photo-gain 146
photogating effect 148, 151
photoluminescence (PL) red-shift 47
photoluminescence spectrum (PL) 119
photon flux 146
photo-thermoelectric effect 147
photovoltaic effect 148, 161
physical vapor deposition (PVD) 85, 110–113, 129, 135, 136, 165
plane-wave DFT 49
plasma-enhanced atomic layer deposition (PEALD) 109
plasmonic effect 148–149
p-n or Schottky junctions 148
point-by-point scanning 91
polarizability 50
polydimethylsiloxane (PDMS) 130
polymethyl-methacrylate (PMMA) 130
polyvinylpyrrolidone solution in dimethylformamide (PVP/DMF) 84
precursors 86, 87, 89, 132, 133, 135–137
pristine graphene 31–33
p-type semiconductor 67, 103
PWmat package 64

q

quantum confinement effect 35, 151
quantum effect 123, 129
QUANTUM ESPRESSO 64
Quasi-Ohmic contact in MoS_2 based field effect transistors 70

r

Raman spectroscopy 115–119, 133
reduced Planck constant 67
relaxed lattice 105
renormalized Fermi velocity 31–33
responsivity 84, 146, 147, 149, 150, 152, 153, 155
$ReS_{2(1-x)}Se_{2x}$ alloy 101, 119
$ReS_{2x}Se_{2(1-x)}$ 101
　monolayer alloy 113
　single-layer alloys 103
reversible elastic tensile strain 41
rise time 146
role of defects in enhanced Fermi level pinning 72–75

s

Sb-doped MoS_2 crystal 121
scanning electron microscope (SEM) 96
scanning transmission electron microscopy (STEM) 93–94, 114, 133
Schottky barrier (SB) 69–72, 75, 109, 149
Schottky–Mott model 69
self-consistent Kohn–Sham potential 62
semiconducting transition metal dichalcogenides (TMDs) 42
semiconductor heterostructures 19, 26, 109, 125
semi-empirical tight-binding method 10–11
Se-rich alloy 103
S-Ga-Ga-S sheet 49
signal to noise ratio (SNR) 126, 147
single-layer Group IV TMDs 104
single-layer $MoSe_2$ 23, 71
single-layer TMDCs materials 126

single-particle Kohn–Sham equation 9
single precursor ammonium thiomolybdate ((NH_4)$_2$MoS$_4$) 86
SiO$_2$/Si substrate 79, 80, 111–113, 133
size control 35–40
slab polarizability 50
Slater–Koster formula 32
SnSe$_{2(1-x)}$S$_{2x}$ 109, 111
SnS$_x$Se$_{1-x}$ alloys 103
Sn$_x$Ge$_{1-x}$S alloys 103
Sn$_x$Ge$_{1-x}$Se alloys 103
solar cells 44, 100, 161
spin-orbital coupling (SOC) 16, 17, 105
stable/diffuse phase diagram 105
state-of-the-art mid-infrared detectors 150
strain energy 16, 21, 26–28
strain engineering 35–51
strain-induced pseudomagnetic field 44
substitutional doping 99
substitution doping 4
supercell geometry 49
supply of reactive deposits 85
switchable optical linear dichroism 51
symmetric ZGNRs 55, 56, 61
symmetry-dependent spin transport properties
 graphene nanoribbons 55–57
 graphyne nanoribbon 57–59

t

thermodynamic fluctuation law 1
thin 2D crystal 85, 86
thin MoS$_2$ films 88
three-dimensional metal electrode 109
three-dimensional stacked graphite 1
tight-binding method 10, 11, 36
time constant 146, 147
T' lattice constants 105
TMDCs/graphene vertical heterojunction 126
TMDCs/TMDCs vertical heterojunctions 126
transition metal chalcogenides (TMCs) 101
transition metal dichalcogenides (TMDCs) 4, 15, 38, 63, 83, 123
transverse acoustic (TA) mode 64
triangular monolayer MoS$_2$ 135
tunnel barrier 69, 70, 150
tunnel diode 126
two-dimensional (2D) atomic sheets 146
two-dimensional (2D) atomic-thick layer 1
2D alloys
 adjustable bandgap 100–103
 carrier type modulation 103–104
 chemical vapor deposition (CVD) 113–114
 chemical vapor transport (CVT) 110–111
 device performance 108–109
 doping of 119–121
 in the field of magnetism 107
 phase change 104–107
 photoluminescence spectrum 119
 physical vapor deposition 111–113
 Raman spectroscopy 115–119
 scanning transmission electron microscopy 114
2D black phosphorus 165
2D doped materials 4, 5, 165
2D GaS nanosheets 49
2D graphyne 37
2D group-VA semiconductors, anisotropic transport properties of 67
2D heterostructures 166
 advantages and application of 125–129
 characterizations of 137–139
 chemical methods 131–137
 conception and categories of 123–125
 mechanical transfer methods 129
2D layered metal oxides/metal hydroxides 4
2D materials
 electric field modulation 48–51

2D materials (contd.)
  perspective of 6–7
  size control 35–40
  strain engineering 40–47
  types of 4–5
  ultra-thin thickness of 165
  wafer-scale and high-crystal quality 165
2D Moiré heterostructures
  graphene/nitrogenated-holey-graphene 26–33
  $MoS_2/MoSe_2$ heterostructures 19
2D MoS_2 38, 48, 153
2D phosphorene 67
2D photodetectors with sensitizers
  graphene based hybrids detectors 151–152
  plasmonic sensitized detectors 153–154
  TMDs based hybrids detectors 152–153
2D $ReSe_2$ 46
2D semiconducting atomic crystals
  density functional theory 9–10
  electronic structure of 11–19
  GW approximation 10
  linear scaling three-dimensional fragment (LS3DF) method 10
  Moiré heterostructures 19
  non-equilibrium Green's function (NEGF) formalism 11
  semi-empirical tight-binding method 10
2D semiconductor-black arsenic 93
2D-semiconductor/metal contacts 72–75
2D semiconductors
  characterization
    band structure 94
    energy dispersive X-ray spectroscopy 95–96
    HRTEM 92–93
    OM 91
    phase structure 93–94
    scanning electron microscope 90
    thickness 92–93

    X-ray photoelectron spectroscopy 94
  electrochemical properties of 96
  optoelectronic memory of 156–161
2D semiconductors preparation
  liquid-phase exfoliation 81–85
  mechanical exfoliation 79–81
  vapor-phase deposition techniques 85–90
2D single materials 4, 165
2D transition metal dichalcogenides (TMDs) 70, 143
two-dimensional graphyne sheet 41
two-dimensional monolayer TMDCs 89
two-dimensional multi-iron materials 129
two-dimensional semiconductor 109
type II heterojunctions 161
typical RCA cleaning method 79

u
uniaxial strain 41, 43
uniaxial tensile/compressive strains 42
uniform strains 41
up-spin density 45

v
valence band maximum (VBM) 35, 64, 75
valence band minimum (VBM) 14, 16
valence band offset (VBO) 17
van der Waal layered three-dimensional materials 4
van der Waals 2D heterostructures 19
van der Waals heterostructure 4, 158–160
van der Waals (vdW) interlayer 69
van der Waals repulsive force interaction 93
vapor-phase deposition techniques 85–90
vapour-phase MoO_3 sulphurization 88
vertical heterostructures 125, 126, 130, 137
viscoelastic stamp 130

W

wafer-scale and high-crystal quality 2D materials   165
wafer-scale growth of 2D materials   6
water freezing-thawing approach   84
waveguide-integrated graphene photodetector   149
weak Van der Waals force   83, 123
WS_2 80
  monolayer   66
WSe_2 80
WSe_2/MoS_2 heterojunction   126
$WSe_{2(1-x)}Te_{2x}$ monolayer alloy   115
WS_2-MoS_2 137
WS_2/MoS_2 few-layer heterojunctions   126
WS_2/MoS_2 heterostuctures   137
$WS_{2x}Se_{2-2x}$ alloy material   101, 113
WTe_2 80
W-Te precursors   137
$W_xMo_{1-x}Te_2$ single-layer alloys   107

X

Xenes   123, 124
X-ray diffraction (XRD)   93, 115
X-ray photoelectron spectroscopy (XPS)   94–95
x–y planar-averaged electron charge difference   25

Z

zero-dimensional fullerenes   1
zigzag α-graphyne nanoribbons   37, 51
zigzag graphene nanoribbon (ZGNR)   35, 55
zigzag MoS_2 nanoribbons   38
zigzag ribbon (z-PNR)   40
zigzag-shaped MX_2 nanoribbons   39
ZrS_2 monolayer   43